Living and Community

Geoff Mulgan

THE EDGE
FUTURES

By the year 2025 the climate will have changed irrevocably, mainly as a result of greenhouse gas emissions. The temperature is predicted to be, on average, half a degree warmer and will fluctuate to a greater extent. Rainfall will have reduced but will also become more extreme. Resources such as energy, water and food imports will be in shorter supply and transport will be constrained; partly as a result of climate change but also due to regulations aimed at preventing global warming. In this series of important and timely books the Edge explore the impact these changes will have on our lives in the future. Global in scope and far reaching in its implications this series examines the significant social, environmental, political, economic and professional challenges that we face in the years ahead.

Contents:

Introduction
Simon Foxell and Adam Poole

If there is one thing in recent decades that has affected the way we live as a community, whether large or small, it is our changing attitude to our time. As we look into the future to approximately 2025 will we have more or less time to offer to living, to our families, to the community?

There is a belief that we work harder and for more hours than previous generations although there is probably little truth in it. Certainly most households, in order to pay the mortgage and supermarket bills, need two, or more, incomes and undirected time for most adults appears to be at a premium or is spent indoors in front of the television and computer.

We are all in a hurry, all under pressure and only interested in communal activity when "there's something in it for me". It is an approach summarised by Margaret Thatcher 20 years ago in her statement to Woman's Own magazine that "There is no such thing as society, only individuals and families."

Government is now keen, possibly desperate, to change this. Policies centre on citizenship skills, participation and engagement and sustainable communities. A champion for volunteering is appointed and the government announces it wants to engage people around the country in a discussion on citizenship and British values (The Governance of the UK, July 2007). But no attempt has been made to find the time when the community will come together, will interact and engage. In 20 years government rhetoric has moved from a denial of society as having an existence let alone a value, to another government begging its citizens to re-engage and to behave as a community again. Where will society be in another 20 years time?

It is frequently suggested that in a world where travel will be prohibitively expensive local communities will find themselves valued again and there will be a return to mutual, neighbourhood support systems. Similarly predicted is the popular discovery that work for purely financial reward is not all that it is cracked up to be, if it does nothing or little for one's quality of life. These arguments fit into current propositions around happiness and mesh with beliefs in slow food and the importance of the convivial lifestyle. They are, of course, a middle class delusion.

We have become used to living in an global world and will not give it up easily, but we may have to get more used to living in it in a virtual sense; enjoying those international products that can come to us without associated transportation costs; images, sounds and even flavours as well as particular types of work and services, rather than the real and physical. Such virtual supping of the world's riches; far-flung friendships based on common interests, computer and communications based work carried out on one continent to satisfy the needs of those on another and a continuing quest for the unusual, stimulating and exotic wherever it is to be found, will not make us more community minded, instead it will encourage greater isolation and inward facing existences.

Religion, often cited as a community based activity that brings a congregation regularly together, has long been heading in this direction. Individuals search for remote guidance and inspiration and exclude those with different sets of beliefs from their social circle. The local and immediate is less engaging than the particular, exclusive and remote. Even those churches, mosques and temples that have thriving congregations find it difficult to want to reach out into the wider, non-spiritual, community. The many community support systems that do exist through religious organisations are frequently holdovers from very different times. New charitable organisations tend to be single-issue based rather than faith generated.

The one exception is in education where faith is uncomfortably bumping up against universal state provision and it is where community-level conflict is most likely to arise. The government, newly interested in community cohesion, needs to think hard how it can make sure that schools are a force for bringing disparate interests together around agreed goods rather than sectarian division.

But, looking ahead, the need to become less-dependant on external resources, whether oil and minerals from politically unacceptable regimes or out-of-season food from the other side of the earth, is likely to become compelling. This inevitably means becoming more reliant on ourselves and finding and rediscovering ways of living within our own means. The ecological footprint of the United Kingdom in 2003 was 5.6 global hectares (gha) as against 1.6 gha bio-capacity, a factor of use vs local supply of 3.5:1. To move towards at ratio of 1:1 is going to require a closer and more local relationship between supply and demand and it is this that will force the new regionalism, so often proposed and discussed, onto our society.[1]

1 Source: Global Footprint Network (www.footprintnetwork.org)

Standard analysis of the economy recognises the public, private and voluntary sectors. Recent history has involved the public sector re-emerging from a nadir of public and government trust and the destructive impulses of the 1980s, but it is now very reliant on private funding, management, resources and leadership even when these have been seen to be lacking. Throughout this period the voluntary sector has quietly trodden a careful path, anxious not to attract the attacks made on the public sector or to over-commit itself when dependent on a fragile and fickle funding stream. Perhaps this sector's time is due when we need to develop stronger communities and a more self-sustaining economy in an equally global but far more energy and resource constrained world.

Time spent in voluntary, community supportive activity will need to be recognised in new ways. This is unlikely to be through salary levels, although the tax system might be adjusted to recognise the social value of time spent not working for profit. Instead it may relate to time made available by both individuals and employers for activities that add to the commonwealth. The education system also needs to prepare students for lives spent not only in a variety of jobs but also contributing their talents in many different ways.

Such activities need to be outward looking and tolerant; building on the strength brought to the UK by its willingness to be a multi-, or even inter-, cultural country. It would be a disaster if by becoming more self-reliant and community focused it also became inward looking rather than internationalist with links and interests maintained across the world. The challenge for the UK is to find ways that are not predicated on consumption and growth but that instead look to stability and providing opportunities for individuals and groups to play an interesting and creative part in a vigorous and communal society.

Managing Communities—Looking to the Future

Geoff Mulgan

Forecasters used to predict that economic growth and technological change would make people ever more detached from the places where they live. Instead, the picture is more complex. Some spaces have been *privatised* and fenced off. But, in other ways, our shared spaces are being democratised and people are becoming more concerned about, and more engaged with, the local environments in which they live and work.

Our work at the Young Foundation stretches back to another period when great struggles were underway over how local spaces should best be organised.[2] In the 1950s Michael Young and his colleagues led part of the assault against the excesses of planning and the movement of people into tower blocks on the edge of London.

2 The Young Foundation was founded in 2005 from the merger of the Institute of Community Studies and the Mutual Aid Centre. The Foundation is a centre for social innovation based in East London—combining practical projects, the creation of new enterprises, research and publishing. Geoff Mulgan is Director of The Young Foundation.

The destruction of social bonds taking place then was the subject of a project and a very influential book published as *Family and Kinship in East London.*[3] That project became, in turn, an inspiration for Jane Jacobs who, a few years later, wrote her seminal work on American cities, *The Death and Life of American Cities*, which was critical in helping people change the way they thought about urban life and communities.[4] Both books encouraged people to look at cities, not as aggregations of buildings and infrastructures, but rather as organic places held together by social and informal bonds that could easily be shattered by ill-conceived plans.

3 Young, Michael and Peter Willmott, *Family and Kinship in East London*, Harmondsworth: Penguin, 1957.

4 Jacobs, Jane, *The Death and Life of American Cities*, New York: Modern Library, 1961.

Since then there has undoubtedly been progress: experts in architecture and planning are now much more likely to recognise and acknowledge some of the features of healthy living communities. Other things being equal; development proposals are more likely to involve the elements which generally encourage a sense of community: reasonable degrees of density, mixed-use; building design which involves lines of sight onto public spaces; connectivity, the mix of both fast flow when you want it, that is to say easy access to public transport, and slow flow when that is appropriate, with places, squares, gardens, corners and parks where people can pause and mingle with others.

But, in other respects, there is still a very long way to go. People are able to exercise very little control over the buildings they live in and the planning and management of their streets. They have been on the receiving end of what has rained down from planners, architects and developers, with little chance to object, and even less chance to get engaged at an early stage of design.

Predicting the Future

Any predictions are risky. I've always liked the comment on one far left leader who, in 1930, said that it was "proof of Trotsky's farsightedness that none of his predictions has yet come true". A famous futurist, Jim Dator, also said that for any prediction about the future to be useful it must at first sound ridiculous. I may have failed by that measure, but I do feel confident about what, for communities, is likely to be the biggest challenge over the next ten to 20 years. It isn't that there won't be enough money, instead it is that society may well become more fragmented; more disconnected and less integrated; with wider gulfs between rich and poor, country and city, religious and secular, and between different races; all resulting in a diminution of social capacity. The causes of fragmentation are many—some are structural, some are consequences of politics. But their common theme is a weakening of the horizontal connections between people, bridging social capital to use the technical term, and the related capacities to empathise, cooperate and get on with others. A recent survey of 11, 13 and 15 year olds in more than 30 countries asked the question "do you find your peers generally kind and helpful?". More than half were able to answer "yes" in every OECD country except the Czech Republic and the United Kingdom where only 43 per cent felt able to answer positively, half the figure in Switzerland and Portugal.

We know the evidence about unwillingness to intervene in street issues and disputes—which shows that UK's citizens are more prone to turn a blind eye than the citizens of other countries. There's no shortage of evidence of people becoming less tolerant, quicker to become angry—whether in the form or road rage or attacks on NHS staff. We know that levels of social trust declined steadily from the 1950s to the 1990s, and although the overall trend appears to have stabilised, it is worrying that 48 per cent of young people aged 11 to 18 years would not trust the 'ordinary man or woman' in the street whereas 30 per cent of adults would. Anti-social behaviour continues to be a top public concern in many areas. When people are asked if life is getting better or worse, a large majority think it is getting worse and the specifics they cite are all about daily interactions, with 47 per cent citing a lack of respect and 46 per cent citing levels of crime.

This isn't about young people and old, though it is sometimes misleadingly presented through this frame. Indeed, in surveys of politeness to tourists in many cities the young tend to score better than the old, and it is men over 60 who come out worst. Instead, it is about how we as humans relate to others, about the civilness of our society: our ability to live together.

So why should any of this matter? It matters because distrust, unfriendliness, rage, a society where people put up shutters, retreat to gated communities and put up internal gates as well, is bound to be a stunted one unable to live up to its potential. It matters too because the presence of a civil society in all its senses is so critical to well-being and happiness. This is one of the messages from the growing mass of evidence on well-being and happiness around the world. What makes societies happy is, in part income, in part good governance. But the evidence again and again reinforces that it is also about trust, about the quality of relationships at the most micro level: how people live together, whether they feel safe walking down their street, talking to a stranger, and whether there is a rough and ready equality of recognition. How spaces are organised, and the messages they send, are vital to how happy we are, and to our ability to thrive. So where do we stand? Who is winning or losing in the battle to shape spaces to meet their needs?

Community Battlegrounds

Childhood

Many British cities are now favoured with much better play spaces than a generation ago: high quality materials and designs, and, in the latest planning codes, much stricter requirements for developers. But, overall civil society has lost ground on issues concerning childhood. Children growing up today are bombarded by an intrusive and often shameless commercial culture and then are taken into the care of the state if things go really wrong, with the voice of civil society barely audible in the din. At the same time many international commentators have observed that cities like London are about as un-child friendly as it is possible to be.

> All families in OECD countries today are aware that childhood is being reshaped by forces whose mainspring is not necessarily the best interests of the child. At the same time, a wide public in the OECD countries is becoming ever more aware that many of the corrosive social problems affecting the quality of life have their genesis in the changing ecology of childhood. Many therefore feel that it is time to attempt to re-gain a degree of understanding, control and direction over what is happening to our children in their most vital, vulnerable years.[5]

UNICEF Innocenti Research Centre

5 Report Card 7 Child poverty in perspective: An overview of child well-being in rich countries, 2007

> The United States (US) and UK ranked at the bottom of a UN survey of child welfare in 21 wealthy countries that assessed subjects from infant mortality to whether children ate dinner with their parents or were bullied at school.[6]

6 *The Washington Post*, 15 February 2007, commenting on Unicef study.

Children have had no direct voice and only a pretty weak indirect say in the key decisions concerning the places they live and play. They are getting more indirect say through regulation on the use of planning gain in development, on the actual quality of play areas, playgrounds and parks, and also some indirect say through things like youth parliaments, youth majors, youth councils and a growing recognition that children should be allowed to speak for themselves and not through an intermediary.

Government has experimented at times with involving children in decisions—for example engaging young teenagers living on estates in interviewing their peers and then presenting recommendations to ministers (at one particularly memorable event, to the Prime Minister and Deputy Prime Minister, where their messages were strikingly different to those coming from the organisations which claimed to represent them). Others have helped shaped provision of leisure facilities. Similar ideas are beginning to have some influence on schools. Children are becoming more involved in design decisions for schools through the Building Schools for the Future programme. This could be a very exciting field for a new style of partnership between planners, architects and others on the one hand and children on the other; rather than seeing them solely as passive beneficiaries.

Old Age and Demography

The second of the battlegrounds is at the other end of the spectrum; adapting spaces, buildings and particularly homes for a very different pattern of old age. A couple, who both reach 65 today, have a 17 per cent chance that one of them will reach 100. Social attitudes have not caught up at all with the extraordinary increase in life expectancy of the UK population, and the building professions, including developers, planners and architects, have been astonishingly slow to grapple with, and design for, new patterns of ageing. These are likely to involve many people moving three or four times after retirement and going through an increasingly more complicated set of life patterns. The baby boomer generation will possibly have ten to 15 years of intense living after retirement; increasingly moving back into city centres, wanting culture, nightlife, fun, food, etc.. A significant proportion of them will have a great deal of disposable income, from a lifetime of well-paid work and especially from downsizing homes, to pay for it.

But, equally, many people will spend much longer periods requiring various degrees of care; for buildings without stairs; and for much more pervasive technology to monitor constantly their health for symptoms that may need attention. Large numbers of the elderly will also require extended terminal care as morbidity, if not mortality, increases.

Philadelphia was earlier this decade the first city in the world to house a family in which five generations lived under one roof. In this decade there will be many more communities and housing models, which either cluster together two, three or four generations. Similarly the work I have been doing in Australia is showing that many people in their 60s and 70s want to live in clusters of perhaps six or seven houses as an alternative to staying in the house they have always lived in, or moving into a care home. There is going to have to be a major set of changes at the interface between building, design, and planning and finance to accommodate these very different patterns of old age. The same will be true in the personal care and health sectors. One of the major requirements will be for the potential and capacity for mutual support in old age, without it being intrusive. Everything we are learning about age is telling us that mutual support is all important, which again takes us back to the excesses in planning and tower block design and the way that they destroyed the capacity for it in the past. It will be a very complex market, both in terms of a need and a desire for different freedoms, products and services as well as the ability to pay for them.

As life-spans increase, hospitals and the healthcare system will have to help a growing population with long term, chronic diseases—from diabetes and MS to heart disease and cancers—look after themselves and those around them. Most of the care provided in the twenty-first century will not come from hospitals, or doctors, or polyclinics—it will be provided by people themselves, and by those around them day in and day out, supported by the NHS, informed by the best knowledge available, and with periodic visits to clinicians. It will require new skills of self-responsibility and cooperation, as well as support networks constructed around the frail elderly or disabled children. It will, in other words, require a much more human-centred, holistic approach that builds on the work of projects like The Expert Patients programme and a wide range of voluntary organisations. But, the critical point is that it will have to be grounded in self-efficacy and social efficacy too, especially if it is to make inroads into inequalities that, in the borough where I work—Tower Hamlets—have left life expectancy 13 years longer at one end of the borough than the other.

Climate Change

The third big shift is going to be not just awareness of climate change and carbon reduction, but the much broader shift to models of sustainable urban and indeed rural living. We can already see this changing the way localities are organised and a city like Havana provides an example of a place transformed by the end of cheap or readily available oil (Soviet oil in Cuba's case); it is a city that is already sourcing the majority of its food from within its own borders.

Havana is a pointer to what a world with far higher carbon prices might look like. It is likely that there would be a revival of urban allotments and small-scale high-quality special production in and around cities; far more encouragement for local sourcing; more visible farmers' markets; organic deliveries, etc., probably alongside the surviving world of supermarkets. There would be far greater use of different travel modes, from electric, rented and shared cars, to smart bicycles and to prominent walking routes in all cities, whether Bogotá, which might already be an exemplar for where we are heading or Los Angeles or even Dubai.

Bogotá is an unusual example of a developing world city. Under two pioneering mayors it has spent huge amounts on cycle paths that run, not just through the city, but right out into the countryside. Of course many Dutch towns, such as Groningen, in one of the richest areas in Europe, have long had a very strong bicycle-orientated transport system. These models of development are very different from the car-based visions of supermarkets, suburbs and ring roads, of which Los Angeles has been the dominant model for the last 60 years, and for which time is running out.

Equally, C40, the grouping of the world's largest cities, all committed to tackling climate change, is discussing the potential shift to neighbourhood energy systems, and energy systems that mainly use urban waste as the source of energy, with combined heat and power (CHP) systems in the localities. This will change not only how we pay for energy but also the psychology of localities, where you are predominantly creating your own energy rather than relying on very big coal-fired power stations or on imported oil. This is another dimension to the re-localisation and recapture of space by communities: create your own rather than the drift to ever-larger scale and ever increasing dependence on big technical systems.

Connectivity and the Sense of Place

The fourth strand is the role of the web and ICT, which is already transforming the sense of locality and place in areas of high connectivity. It has turned out to be a far more interesting impact than was feared ten or 20 years ago. It is becoming, in many different ways, a tool for the re-assertion of locality and for the particularity and ownership of place.

Technology is beginning to create new communities much more directly. There has been experimentation in some countries along the lines of giving everyone in the street the email addresses of everyone else in the street or by establishing neighbourhood web sites, exchange systems and/or neighbourhood news services. In almost every case, the effect of these IT initiatives has been to increase the amount of face-to-face contact people have with their neighbours; knowing about their neighbours has encouraged the feeling that new arrivals were part of the community too. Far from the internet and communication technologies pulling people away from place as was warned and feared, the technology has been able to simultaneously establish local as well as global bonds.

Two Examples

One is a project, which The Young Foundation launched earlier this year with MySociety and which won a prize for the best new civic media. Called Fix My Street, it is a simple website using Google maps.[7] Any citizen can report a broken piece of civic infrastructure, an abandoned car or anything like that. On the website, enter the postcode, click on the interactive map and send a message to the responsible person in the council whose job it is to fix it. This is done publicly so that anyone else can comment and see whether the problem is fixed promptly or not.

7 www.fixmystreet.com

Sites like Fix My Street are making space more visible and collectively owned than it was, or could have been, in a pre-web era. There are now people working on a version of this called Fix My Planet, which again uses Google maps to identify particularly high emissions from buildings, factories and so on and, as has happened with the Clean India project, in India, it is mobilising teenagers as the guardians of ecology space.

Another example is one for which we have so far only developed a prototype. Called www.yourhistoryhere, it is a site that is intended to develop and create new membranes for the city. If you walk along a street in East London, in Glasgow or Manchester; it is very difficult to get any sense of what happened there in the past; what information or memories exist about that place; whether any significant events took place or people lived or worked there: from battles to local celebrities. There are various routes into a city's past, through books in libraries and so on, but they are almost inaccessible and very slow to interrogate.

Again, the very simple idea is to use Google maps or its successors and layer a series of membranes over the base map which can then be used to find out about a place, its past and significance. Some of these membranes will just be writings on events that happened in a particular house, street or park. It may have been a great demonstration, or maybe a novel was written there or a crucial meeting held in an upstairs room. Such pieces of information can be layered over recordings of oral history; people saying what it was like walking down the street in the 1920s or in the 1940s, as the bombs fell during the Blitz; or even a piece of film. The idea is that you steadily build up, layer upon layer, information about places using the web and, in this way, make the past very immediate and, ultimately accessible via a mobile phone or a handheld device, so that as you are walking down Princess Street in Edinburgh, if you want, you can talk to the buildings. It means the city, which is normally mute in terms of its meanings, its passions and its history, becomes vocal and is brought alive by technology.

Technology has the capacity to re-awaken the local and make places meaningful. We thought this was particularly important for new immigrants, and for people coming to an area such as the East End, from Somalia or Bangladesh, who, at the moment, have no way of knowing about the events and history of the area, what is meaningful, what happened in the past and why they matter to others. I think we are going to see an extraordinary explosion in the next ten to 20 years of innovation around the technology of place—based initially around Google maps and Google Earth but potentially going off in many other different directions as well.

Policing

In many ways British policing leads the world. The experience of rioting in the UK in the 1980s, in Brixton and in Toxteth and in the northwest in 2001 prompted a fundamental rethink and the start of micro-policing—an attention to small details to avoid the snowballing effects of bad policing. The police have recognised that their success depends on working at a much more local level, getting to know neighbourhoods and being accountable to very local areas. This trend—which in its recent incarnation started in Tower Hamlets in London—has reinforced the growth of new roles (in particular the Police Community Support Officer scheme) and new styles of policing (such as the return to bicycles). The experts were often sceptical about having more accessible and visible police. But this shift to a more community-based model of policing is undoubtedly one of the reasons not only why crime has fallen so remarkably over the last 15 years (by at least a third) but also why fear of crime has fallen despite the best efforts of the media to convince people that crime is ubiquitous. It is also, of course, one of the reasons why, despite severe social dislocations, the UK has not seen riots equivalent to those that have repeatedly swept French cities in recent years.

Well-being

Over the next few years, the subject of well-being is going to reach the mainstream agenda. Many local areas are also thinking hard about how they can improve well-being, rather than focusing solely on more traditional measures of success, such as jobs and school performance. Work is underway in many cities around the world to define new metrics of well-being, with active engagement from the OECD amongst others. The Young Foundation is working in three areas (Manchester, Hertfordshire and South Tyneside) on an ambitious programme of work to test out what really has an impact on well-being. One strand involves young people learning how to be resilient; another is aimed at isolated elderly people; and a third on parenting. Several of the strands are very much about place—including one exploring the data which shows that happiness levels correlate with how well you know your neighbours, and another seeking environmentally useful actions that also make people feel good about themselves and their communities.

Governance

To make the most of these many ideas we need to rethink governance. The irony of local government is not just that it does not govern much: it is also that it is not very local. Our lowest tier of local government is still very distant from most people, with an average size of about 115,000 people, compared to more like 10,000 in most Western countries. One consequence is that the UK has one elected representative for every 3,500 citizens whereas France has one for every 100 people. There almost everyone knows someone who is involved in government and representatives are closely involved in their communities.

While the UK has centralised, almost every other major country has gone in the opposite direction. Countries as varied as Italy, Spain, France, India, China and Brazil have been passing power downwards. The argument used to justify the UK's peculiar stance was that centralisation would deliver better services and better results. Whitehall, we are told, is simply more efficient than town halls. This argument looks less credible in the wake of lost data and the multiple cock-ups around migration, and it is even less credible when you look at the facts: the most recent surveys of public service performance show the UK bumping along at the bottom with the US, while countries with much more decentralised systems are well ahead on measurable outcomes.

Many of the big trends which are likely to shape the next few decades point in a localist direction. Climate change is encouraging people to think again about sourcing local food, working locally, driving less and walking more. Equally, an ageing population is likely to care more about the local quality of life. Even the internet is, paradoxically, doing much to strengthen local ties as people find new ways to link up with others living near them.

For all of these reasons the time is ripe for a turn against centralisation, and for passing power not just from national government to local authorities, but also from local councils right down to neighbourhoods. This is where democracy needs to start, ideally with directly elected neighbourhood councils, a modernised version of existing parish councils, which should be responsible for issues such as public spaces and play areas. Modest annual precepts (for example £20 a year) would provide significant enough budgets to get a lot done. I would encourage these neighbourhood councils also to have formal influence over the council when it is debating issues that affect the area, for example parking policies.

But the top priority is to establish institutions with the power to fix the day-to-day problems that are so often most infuriating to residents.

We then need to re-empower local government itself. This isn't something that can be done quickly. Half a century ago the most energetic and able people in the community would automatically think of standing for public office. Now the average age of councillors is 58 and most ambitious politicians want to go onto the backbenches in Westminster, not to prove themselves running a town or a city or a county. It will take a long time to turn that around. But, as councils regain the power to make real decisions, they will also attract more people to stand. Money is critical to this, and although government has dithered I'm convinced that we will see some control over taxation and spending pass back to local government, starting off with the relatively marginal taxes like business rates and taxes on development, but, in time, moving onto the big ticket items, income tax and VAT. We are already seeing a reversal of centralisation in inspections and targets. I doubt national targets will ever disappear entirely, but they are being made more flexible and more responsive to specific locations' needs. One reason for keeping some is that external pressure can improve performance. There are many fewer truly dire councils than a decade or two ago partly because of the pressure from inspections. Local government needs to be challenged from above as well as from below, just as national governments benefit from the challenge they occasionally get from the European Commission or the OECD.

Not a Broken Society

One of the worst things that has happened to many communities around the world is when high levels of crime, often associated with drugs, lead people to turn inwards and distrust those around them. In the late 1980s/early

1990s it was almost possible to correlate by class how much people talked to their neighbours, but, also, in reverse correlation, how much they trusted their neighbours. In poorer areas, where people had no choice but to interact with those around them, they were becoming less and less trusting of their neighbours, whereas in more prosperous areas, where people did not interact that much, they felt comfortable and far more trusting. This reversed the social cohesiveness of the 1950s and earlier, when the poorer and more deprived areas would have had the much tighter social bond.

But over the last decade most of the evidence suggests that these trends have gone into reverse. Research on poorer neighbourhoods shows that most have improved with significantly lower crime, more jobs and better health outcomes. Indeed in research done by the LSE there is a stark comparison between the UK and the US and others. Where poor US neighbourhoods have remained poor and often declined, even while the cities around them have enjoyed the long economic boom, the British ones have generally improved. A large part of the reason is that the British state has continued to be active, providing services, healthcare and schools, as well as investing in regeneration.

Yet, in the last year David Cameron, Ian Duncan Smith and the Archbishop of Canterbury have all, in a single week, described the UK as a 'broken' society. Readers of newspapers could easily believe this—yet the claim flies in the face of the evidence. Social capital did fall from the 1950s through to the 1990s, but it appears that it has been on the rise again in the last ten years. Not by a large amount, admittedly, but perhaps this is not surprising since this is also the period that has seen the largest fall in crime in the century. Surveys also show that people feel more comfortable in their communities since their willingness to help their neighbours appears to be going up.

There are undoubtedly many things that are getting much worse rather than better, but, on the whole, it is very hard to claim that the UK is a more broken society than at any other time. If you look, for example, at generational relationships, they were far worse in the 1960s and 70s—with much more distrust and disconnection between generations—than there is now. If you look at distrust between the races and at levels of racism, again, it is substantially less than a generation or two ago.

Communities also remain quite strong and people are naturally quite helpful to each other. A Mori survey last year for The Young Foundation looked at where people turn for help. They looked at a range of situations from help in the garden all the way to dealing with a serious illness. It confirmed that, overwhelmingly, the important sources of support are still friends and family, with family being far and away the most important. The market and the state are much less important in people's everyday lives; organised religion is almost invisible.

This strength can also be seen in the remarkable resilience of British communities. The past decade has brought a phenomenal number of migrants to the UK (and to mention just one figure, over 27 per cent Londoners recorded by the 2001 Census, were born outside the UK).[8] Yet, so far, the response has been calm—no widespread riots, no dramatic swing to racist parties. Even infrastructure systems, which appear stretched, are, in fact, against expectations, functioning perfectly adequately. That is not to say that there aren't big challenges—and a challenge more for London than for other places is how to cope with this great fluidity and high turnover of people.

8 Key Facts for Diverse Communities: Ethnicity and Faith, Greater London Authority, Data Management and Analysis Group, 2007

How fears are talked about matters because it can lead to very different responses. The fears of strangers in recent years have encouraged largely technical solutions. The UK is exceptional worldwide for its 4.3 million CCTV cameras and for the fact that the population is relatively relaxed about this intrusion into their lives. The phenomenal increase in video surveillance, in speed cameras, the commercial use of personal identities and a community's CCTV being made available for communities to self-monitor—examples include the police publishing photographs on the web of curb crawlers or people dealing in drugs outside a tube station—is changing the social dynamics and character of a place, with all the delicate interdependencies that sets in motion. Public spaces were once quite private—now they are not.

The alternative, which uses the presence of people to reduce fears, have been sidelined and this has happened during a period when many roles—park keepers, station wardens—have been cut back. But experience suggests that the presence of people is a better way of making spaces safe: a good flow of people, plenty of eyes and some official or semi-official roles responsible for making spaces work (like town centre managers or estate concierges).

Belonging

The study *Family and Kinship in East London*, mentioned earlier, portrayed a set of very dense ties of belonging: both to place and to people; mainly through matriarchs, strong women who had held the community together. When it was published, the study was an argument against the dispersal of communities to Essex and destroying the things that made the community work.

> The mothers represent tradition. They hold to religion and to the old ways more tenaciously than their children, and may be up against the more modern ideas learnt by the wives, and even more by the husbands, from sources outside the family.[9]

9 Young, Michael and Peter Willmott, *Family and Kinship in East London*, 1957, Harmondsworth: Penguin, p. 56

40 years later, some of the same team, including Michael Young, went back to the same streets to see what had happened, and discovered that many of the same patterns of living and community could now be observed in the Bangladeshi population living in the same streets the former white working class had lived in the 1950s and who were the subject of the first study.[10] They found very similar dense networks of mutual support, with women playing pivotal roles, although possibly different in nature, and with the street remaining a very live place of interaction.

10 Dench, Geoff, Gavron Kate and Young Michael, *The New East End: Kinship*, Race and Conflict, 2006, Profile

Back in the nineteenth century, a lot of people were moved out of the British countryside and away from very stable communities. They were uprooted and put into cities which were as close to visions of hell as one could imagine—utterly atomised, utterly anomic, with very high levels of crime and disease and appalling poverty. And yet, within a generation or two, they had become settled and developed a fairly strong sense of community, so that by the middle of the twentieth century they thought of themselves as having always been in East Manchester or East London, or wherever, with these places being seen as exemplars of strong communal life.

In recent years we have been through another phase of dislocation and de-urbanisation, with the loss of jobs with, in some cases, de-population being followed by the arrival of a large migrant community. In areas such as East London, it is very striking how people have lost the sense of belonging they once had. If they and their family have lived for a long time in an area that is now full of newcomers they may feel they have lost their place, their icons and their memories. But, equally, many of the new migrants do not feel they really belong there either. The Young Foundation, with its East End roots, has been focusing on these issues. We've been trying to understand better what makes a person, be he/she a newcomer or a long-term inhabitant feel he/she belongs, or, alternatively, does not belong in a place. The Foundation has been carrying out research in Barking and Stoke, in part trying to understand support for the British National Party through the lens of belonging, and, out of it I think we have come up with a very simple way of thinking about this, which, if it is right, is very different from some of the other theories in circulation, such as the theory of social capital put forward by Robert Putnam in *Bowling Alone*, 1995.

Whether you feel or believe you belong depends in part on the messages you receive from your immediate environment. Human beings are intrinsically programmed, for clear evolutionary reasons, to tell whether they fit in and are in an environment that will help them to survive or not. We all are very adept at picking up cues from the people around us as to whether we belong or are welcome in a particular place or situation. Building from that very simple starting point, and thinking about an individual in a definite location—a Somali in Tower Hamlets or a white former car worker in Dagenham perhaps—we have come up with an approach that looks at the feedback people get from a place and the subliminal messages that place sends you about your position in it.

Some of these messages will be social: about the presence of friends and family to look after you in times of crisis. Some will be economic, telling you if this place will provide you with a living and with a job, with some recognition of your value in having something to contribute. If you are unemployed or face discrimination there will be cultural messages relating to whether you see your way of life reflected in the official culture, events and festivals, or not. The physical shape of places communicates safety or threat. There will be messages about politics; if you see people like yourself in positions of power, or in a position of authority in religion or in the market. There will also be the messages you get from your fellow citizens—whether they look at you with hostility and distrust, or whether you feel safe and indeed, whether you are in fact safe.

The sense of belonging is usually a rational response to the feedback we get from both the physical and the human environment around us. The great challenge for many cities going through rapid change is in being able to adjust these feedback messages to suit new circumstances. Many white working class communities, not just in London but also in other parts of the UK, are getting very negative feedback messages: messages that their skills are not wanted, they are not wanted in jobs and their culture is worthless. These messages are often based on factual misinformation about such things as housing allocations, but, nevertheless, they are important and are fuelling things such as the rise of the BNP. Equally, for many of the migrant groups coming into the UK, particularly the ones arriving in relatively small numbers, the UK is very bad at making them feel welcome and at providing a route to achieving of a sense of belonging. We can and should learn lessons from places such as Canada, which holds dinners and receptions for new migrants and has a whole civic movement devoted to welcoming and integration. There are some good small-scale examples of integration, for example in Tottenham, where there is a 'meet the neighbours' programme, where each community lays on meals and events to explain itself to the wider community. But, overall, such initiatives are still rare and the cultivation of belonging has not been a priority for councils, developers, housing associations or anyone else.

This, as I have said, runs counter to what Robert Putnam has proposed. He has argued that where you have more diverse areas you will get lower social trust and lower capital. Our model indicates that you can have very diverse social areas but with the different groups all getting strong feedback through a range of sources and routes, validating and recognising them as belonging. Finchley would be a good example. It used to be Mrs Thatcher's constituency, but has since become a very diverse part of North London, with most people having friends and families in the area and able to get jobs in the local economy. It is a reasonably open political system that leaves relatively few feeling alienated or excluded. You could compare this to an all white housing estate in Sunderland that feels completely excluded by everything else which is going on around it. Such an analysis can suggest ways for any particular area to deal with causes of alienation and exclusion that otherwise might tear it apart.

The power of a simple model like this comes down to what is at the heart of this book, which is how people and environments relate to each other. The messages that people receive from their environment; whether they fit in, are valued and have a place, or whether it is the other way around, and they are forced to exist in, and submit to, an essentially hostile environment.

The connecting thread behind all of this is that most people now expect to have much more say over many different dimensions of their lives than they did a generation or two ago. They expect it in terms of their identities, what happens in their kitchen, in their bedroom or in relation to politics. But how we organise physical space and make communities work and live has, in many respects, lagged behind. It is a sector dominated by very powerful architects, developers and planners, who have learnt how to pay lip service to the public but are not so good at providing what is necessary and desired in practice.

What is the challenge for the next 15 years?
To shape places to fit human needs we need to start with some notion of what is likely to happen over the next 30 years to make sure we are prepared for it. For example, given what we know about Climate Change over the next few decades we need to re-engineer both our physical and supply infrastructure so that housing will remain comfortable, can cope with more extreme weather conditions and will be, able to run on close to zero carbon emissions. We need to make transport systems that are less car dependent, have energy systems that are based on neighbourhood networks; source more of our food locally and much else.

It is undoubtedly better to be ahead of the curve rather than behind it if you want your communities still to be successful in 20 or 30 years' time. It is almost certain that there will be a larger elderly population with many more people coping with chronic disease. This alone has some very straightforward implications for the physical design of buildings and for mobility in towns and cities. Each and every competent place shaper needs to be thinking about these issues now. But out of the belonging argument will also come some subtler and more difficult tasks about how you cultivate long-lasting and resilient communities. Some of these concern economics; to ensure widespread access to the mainstream economy and to develop jobs and the skills required to fill them across most if not all social groupings. The state may have to intervene to ensure that disadvantaged groups have access to the informal networks of the type that support society and that power structures, whether they are politicians or the police, reflect the make-up of the community so that people feel as if they belong.

It is also important to know who makes places work. The Young Foundation has developed a tool called the Social Network Analysis Method. It is an internet-based questionnaire that helps build up a picture of who is influential and from what position. The analysis starts at the level of a town or district. In any one field, be it crime reduction or physical development, there will be dozens or hundreds of people working; some in local government; some in other agencies and some in the private sector. The information we acquire is used to map who helps whom and who provides information. It is a way of mapping collaboration within a community that also tells you who and where the blockers are.

We have also, last year, been using the method in King's Lynn, studying the local community from the bottom-up, to understand who are the people who make things happen in a low income neighbourhood. It should eventually become one of the mainstream methods for understanding the real social dynamic of places, a very different perspective from the classic diagrammatic and hierarchical view of organisations and very useful for patching in the most effective intelligence and communication methods for the locality.

If the question is how, with a community focus, do you prepare, over a ten to 20 year period, to tackle the big challenges ahead, be they radically halving carbon use, significantly reducing the percentage of chronic illness in the population, working with a significantly more diverse population that has a 25 per cent and not 90 per cent minority then there are not off-the-shelf answers. The way forward is make sure, as a nation, that places are experimenting with a lot of different models and that we are all collectively learning as quickly as possible which ones work and which ones do not. With this evolutionary approach, we experiment with neighbourhood energy systems of different kinds and with neighbourhood waste and we experiment with different kinds of housing models. We find what works, share the knowledge and roll out an improved approach.

While it is a mainstream approach for science it is harder to apply in politics. Roosevelt famously did it during the Depression: dealing with mass unemployment, he said of "course I am going to try anything and if some things fail I will try the next thing. What else do you expect me to do?". It is an approach that, curiously, has also been applied in The Building Schools for the Future programme. I say curiously because the timescale of the BSF programme makes it slightly flawed since you can never discover what it is that works in time, because it takes three years to build a new school and then four years to know what is really working.

I am possibly over-confident on climate change. When I worked in government I oversaw the UK's strategy of cutting carbon emissions by 60 per cent by the year 2050. Although few of the recommendations were politically acceptable then, in the 1990s, or in 2001, I have been surprised by how quickly some of these ideas and potential policies, which were then off limits, have started to become acceptable, at least at a conversational level. The pace of change in attitudes is fast and could become even faster, allowing robust and hard-hitting policies to be adopted without disaster across Europe.

What needs to happen in order to shift to a low carbon economy is an acceleration of trends, not a dramatic reversal of trends. And, as long as we have enough time that is fine. Many similar shifts have happened in the past, and with the necessary rapidity, such as during with the energy crisis in the 1970s and even the recent de-materialisation of various parts of commercial activity into virtual economic space. The real question is whether we think we have a 50 year or a ten year transition. With a 50 year transition, it is a straight-forward problem, but, if it is a ten or a 20 year transition then it is extraordinarily challenging by any historical precedent.

Energy systems take many, many decades to turn around and, in that respect, Climate Change is like urban change. It is the interaction of three types of change process.

1. Top down—command and control: the world of laws, regulation and post-Kyoto treaties, which will order us to have energy efficient light bulbs and will ban certain categories of car.
2. The horizontal pressures of markets responding to uncertainties and investing in problem solving and efficiency gaining technologies; for example low-energy processes or zero waste procedures, in competition between businesses.
3. Bottom-up pressure from communities and children, which, so far, has made much of the running on life-style change and changing approaches to food and moves towards such things as neighbourhood energy systems.

Real progress is likely to depend very much on bottom-up pull as well as top-down push. Children in particular are becoming very powerful intergenerational change agents.

I think this may be much less difficult than say the first wave of Industrial Revolution capitalism, where it then took many many decades to establish a social contract, so that model of society did not destroy humanity with child labour and slavery and appalling cities. Over time, people have negotiated a better balance with humanity—and are still trying to do so. We are probably 40 years into a similar story with the environment, of trying to ensure that we now have a model for the economy that does not utterly jeopardise nature and we have 40 to 50 years of experience of some of the ways of doing it. I do not see why it will be inherently harder now to undertake something similar to what we have already achieved in society, and, remember, we also had then the innumerable experts who said you could not ban child labour or slavery or introduce a welfare state or a health system without destroying economic growth. Exactly the same arguments are now made about climate change.

It is in places that we can see both the dangers of losing our belief in the possibility of shaping our own destinies and the opportunities. The worst fate for any place is to become fatalistic—to believe that there is nothing to be done. Modern politics grew out of localities—and the experience of improvement, public health, schooling, reducing crime, welfare, mutual help in the nineteenth century all paved the way for the confident democratic politics in the twentieth century. But the confident democratic politics, as we have known them have not been enough: again and again places have lost that sense of destiny and freedom and become victims of planners, developers, and global forces.

Places, however, can be autonomous but that requires a politics that is brought closer to home and it requires that citizens take responsibility for the world around them, rather than drifting into an angry but passive resentment. Today, that autonomy comes in the context of radically greater interdependence—of lives interwoven through the economy, flows of people and information. This is what drives so many of the most important movements of our time, from fair trade to Slow Food. Yet too few of our society's moral thinkers and even fewer of our institutions have adequately adapted to this fact. Hopefully the next generation of places will truly demonstrate how you can be simultaneously local and global, strongly connected to those around you but also open to your place in the world.

Suggested Reading

General

"Barker Review of Land Use Planning", Department of Communities and Local Government, 2006.

Florida, Richard, *The Rise of the Creative Class*, London: Basic Books, 2002.

Foxell Simon, ed., *The professionals' choice: The future of the built environment professions*, London: Building Futures, 2003.

Kunstler, James Howard, *The Long Emergency*, Atlantic Monthly Press, 2005.

Leadbeater, Charles, *Personalisation through participation: A new script for public services*, London: Demos, 2004.

Schumacher, EF, *Small is Beautiful*, Vancouver: Hartley & Marks,1999.

Economic Survey of the United Kingdom, OECD, 2007.
World Population Prospects: The 2006 Revision, Population Division of the Department of Economic and Social Affairs of the United Nations Secretariat, United Nations, 2007.

Planet Earth and Climate Change

Flannery, Tim, *The Weather Makers: The History and Future Impact of Climate Change*, Melbourne: Text Publishing, 2005.

Gore, Al, *Earth in the Balance: Ecology and the Human Spirit*, Boston: Houghton Mifflin, 1992.

Gore, Al, *The Assault on Reason*, Harmondsworth: Penguin, 2007.

Hartmann, Thom, *Last Hours of Ancient Sunlight*, New York: Three Rivers Press, 1997 (rev. 2004).

Hawken, Lovins & Lovins, *Natural Capitalism*, London: Little Brown, 1999.

Hillman, *Mayer, How We Can Save the Planet*, Harmondsworth: Penguin, 2004.

Homer-Dixon, Thomas, *The Upside of Down: Catastrophe, Creativity and the Renewal of Civilisation*, New York: Alfred A Knopf, 2006.

Kolbert, Elizabeth, *Field Notes from a Catastrophe: A Frontline Report on Climate Change*, London: Bloomsbury, 2006.

Lovelock, James, *Gaia: A New Look at Life on Earth*, Oxford: Oxford University Press, 1979.

Lovelock, James, *The Revenge of Gaia*, London: Allen Lane, 2006.

Lynas, Mark, *High Tide: The Truth About Our Climate Crisis*, London: Picador, 2004.

Lynas, Mark, *Six Degrees: Our Future on a Hotter Planet*, London: Fourth Estate, 2007.

Marshall, George, *Carbon Detox*, London: Gaia Thinking, 2007.

McDonough, W, and Braungert M, *Cradle to Cradle, Remaking the Way*

We Make Things, New York: North Point Press, 2002.

Monbiot, George, *Heat: How We Can Stop the Planet Burning,* London: Allen Lane, 2006.

Walker, G, and King D, *The Hot Topic: How to Tackle Global Warming and Still Keep the Lights On*, London: Bloomsbury, 2008.

Action Today to Protect Tomorrow—The Mayor's Climate Change Action Plan, London: GLA, 2007.

Climate Change The UK Programme, London: DEFRA, 2006.

Summary for Policymakers of the Synthesis Report of the IPCC Fourth Assessment Report, United Nations, 2007.

Cities

Girardet, Herbert, *Cities People Planet: Liveable Cities for a Sustainable World*, Chichester: Wiley-Academy, 2004.

Jacobs, Jane, *The Death and Life of Great American Cities*, New York: Random House, 1961.

Jacobs, Jane, *The Economy of Cities*, New York: Random House, 1969.

Mumford, Lewis, *The Culture of Cities*, New York: Secker & Warburg, 1938.

Sudjic, Deyan, "Cities on the edge of chaos", *The Observer*, March 2008.

Urban Task Force, *Towards an Urban Renaissance*, London: E&FN Spon, 1999.

Work

Abramson, Daniel M, *Building the Bank of England*, New Haven, CT: Yale University Press, 2005.

Alexander, Christopher, *The Timeless Way of Building*, Oxford: Oxford University Press, 1979.

Anderson, Ray, *Mid-Course Correction: The Interface Model*, Chelsea Green, 2007.

Brand, Stewart, *How Buildings Learn*, New York: Viking Press, 1994.

Brinkley, Ian, *Defining the Knowledge Economy, Knowledge Economy Programme Report*, London: The Work Foundation, 2006.

Castells, Manuel, *The Information Age: Economy, Society, Culture*, Oxford: Blackwell, 1996.

Davenport, *Tom, Thinking for a Living*, Boston: Harvard Business School Press, 2005.

Dodgson, Gann, and Salter, *Think, Play, Do*, Oxford, 2005.

Dodgson, Gann and Salter, *The management of technological innovation strategy and practice*, Oxford: Oxford University Press, 2008.

Duffy, Francis, *The Changing Workplace*, London: Phaidon, 1992.

Duffy, Francis, *The New Office*, London: Conran Octopus, 1997.

Duffy, Francis, *Architectural Knowledge*, London: E&FN Spon, 1998.

Duffy, Cave, Worthington, *Planning Office Space*, London: The Architectural Press, 1976.

Galloway, L, *Office Management: Its Principles and Practice*, Oxford: The Ronald Press, 1918.

Gann, David, *Building Innovation*, London: Thomas Telford, 2000.

Giedion, Siegfried, *Mechanization Takes Command*, Oxford: Oxford University Press, 1948.

Gilbreth, FB, *Motion Study*, New York: Van Nostrand, 1911.

Gottfried, David, *Greed to Green*, Berkeley, CA: Worldbuild Publishing, 2004.

Groak, Steven, *Is Construction an Industry?*, Construction Management and Economics, 1994.

Handy, Charles, *Understanding Organizations*, Harmondsworth: Penguin, 1967.

Hawken, Paul, *The Ecology of Commerce*, New York: HarperCollins, 1993.

Mitchell, William J, *City of Bits*, Cambridge, MA: MIT Press, 1995.

Quinan, Jack, *Frank Lloyd Wright's Larkin Building*, Cambridge, MA: MIT Press, 1987.

Sassen, Saskia, *A Sociology of Globalization*, New York: Norton, 2006.

Sennett, Richard, *The Culture of the New Capitalism*, New Haven, CT: Yale University Press, 2006.

Taylor, Frederick, *The Principles of Scientific Management*, New York: Harper & Brothers, 1911.

Trease, Geoffrey, *Samuel Pepys and His World*, London: Thames and Hudson, 1972.

Education

Aston and Bekhradnia, *Demand for Graduates: A review of the economic evidence*, Higher Education Policy Institute, 2003.

Friere, Paolo, *Education: the practice of freedom*, London: Writers and Readers Cooperative, 1974.

Gardner, Howard, *Multiple Intelligences*, New York: Basic Books, 1993.

Goodman, Paul, *Growing up absurd*, New York: First Sphere Books, 1970.

Illich, Ivan, *Deschooling Society*, London: Calder and Boyars.1971.

Kimber, Mike, Does Size Matter? *Distributed leadership in small secondary schools*, National College for School Leadership, 2003.

Nair and Fielding, *The Language of School Design*, DesignShare, 2005.

Neil, AS, Summerhill, Harmondsworth: Penguin Books,1968. *The Children's Plan—Building Brighter Futures*, DCSF, December 2007.

Every Child Matters: Change for Children, DfES/HM Government, 2004.

Higher Standards, Better Schools For All, DfES.

2020 Vision Report of the Teaching and Learning in 2020, Review Group, 2006.

www.smallschools.org.uk

www.thecademy.net/inclusiontrust.org/Welcome.html

www.eco-schools.org.uk

www.standards.dfes.gov.uk/personalisedlearning/about/

Transport and Neighbourhoods

Banister, David, *Unsustainable Transport: City Transport in the New Century*, London: E&FN Spon, 2005.

Bertolini L, and T, Spit, *Cities on Rails. The Redevelopment of Railway Station Areas*, London: Spon/Routledge, 1998.

Calthorpe P, and Fulton, W, *The Regional City: Planning for the End of Sprawl*, Washington, DC: Island Press, 2003.

Dittmar H, and Ohland, G, *The New Transit Town: Best Practices in Transit-Oriented Development*, Washington, DC: Island Press, 2004.

Hickman, R and Banister, D, *Looking over the horizon, Transport and reduced CO_2 emissions in the UK by 2030*, Transport Policy, 2007.

Holtzclaw, Clear, Dittmar, Goldstein and Haas, *Location Efficiency: Neighborhood and Socioeconomic Characteristics Determine Auto Ownership and Use*, Transportation Planning and Technology (Vol. 25) 2002.

Commission for Integrated Transport, Planning for High Speed Rail Needed Now, 2004, viewed at http://www.cfit.gov.uk/pn/040209/index.htm

Regional Transport Statistics, National Statistics and Department for Transport, 2006 Edition.

Energy, Transport and Environment Indicators, Eurostat, 2005 Edition.

Toward a Sustainable Transport system, Department for Transport, 2007.

Eddington Transport Study, HM Treasury & Department for Transport, 2007.

UK Foresight programme, *Tackling Obesities: Future Choices*, The Government Office for Science and Technology, 2007.

Community

Dench G, Gavron K, and Young M, *The New East End: Kinship*, Race and Conflict, London: Profile, 2006.

Jacobs, Jane, *The Death and Life of American Cities*, New York: Modern Library, 1961.

Putnam, Robert, *Bowling Alone: The Collapse and Revival of American Community*, New York: Simon & Schuster, 2000.

Young M, and Willmott, P, *Family and Kinship in East London*, Harmondsworth: Penguin, 1957.

Report Card 7, *Child poverty in perspective: An overview of child well-being in rich countries*, UNICEF Innocenti Research Centre, 2007.

Key Facts for Diverse Communities: Ethnicity and Faith, Greater London Authority, Data Management and Analysis Group, 2007.

www.footprintnetwork.org

www.yourhistoryhere

www.fixmystreet.com

Globalisation

Abbott, C, Rogers, P, Sloboda, J, *Global Responses to Global Threats: Sustainable Security for the 21st Century*, Oxford: The Oxford Research Group, 2006.

Balls E, Healey J and Leslie C, *Evolution and Devolution in England*, New Local Government Network, 2006.

Gladwell, Malcolm, *The Tipping Point: How Little Things Can Make a Big Difference*, London: Little Brown, 2000.

Goldsmith, Edward, "How to Feed People under a Regime of Climate Change", *Ecologist Magazine*, 2004.

Gore, Al, *The Assault on Reason*, London: Bloomsbury, 2007.

Gray, John, *Black Mass: Apocalyptic Religion and the Death of Utopia*, London: Allen Lane, 2007.

Guillebaud, John, Youthquake: *Population, Fertility and Environment in the 21st Century*, Optimum Population Trust, 2007.

Hines, Colin, *Localisation: A Global Manifesto*, London: Earthscan, 2000.

Kagan, Robert, *Of Paradise and Power: America and Europe in the New World Order*, New York: Alfred Knopf, 2003.

Martin, James, *The Meaning of the 21st Century*, London: Transworld, 2007.

Meadows, Meadows, Randers and Behrens, *Limits to Growth*, Club of Rome, 1972.

Nordhaus, T, and M, Shellenberger, *Break Through: From the Death of Environmentalism to the Politics of Possibility*, Boston: Houghton Mifflin, 2007.

Porritt, Jonathon, *Capitalism: As if the World Matters*, London: Earthscan, 2005.

Roszak, Theodore, *World Beware! American Triumphalism in an Age of Terror*, Toronto: Between the Lines, 2006.

Sachs, W, and T, Santarius *Fair Future: Resource Conflicts, Security and Global Justice*, London: Zed Books, 2005.

Kirkpatrick Sale, *Dwellers in the Land*, New Society Publishers, 1991.

Shrybman, Steven, *A Citizen's Guide to the World Trade Organisation*, Ottawa, Canadian Center for Policy Alternatives, 1999.

Soros, George, *The Age of Fallability: The Consequences of the War on Terror*, Beverly Hills, CA: Phoenix Books, 2006.

Stern, Nicholas, *The Economics of Climate Change: The Stern Review*, Cambridge: Cambridge University Press, 2007.

Stiglitz, Joseph, *Globalization and its Discontents*, New York: Norton, 2002.

Stiglitz, Joseph, *Making Globalization Work*, New York: Norton, 2006.

Wolf, Martin, *Why Globalization Works*, New Haven, CT: Yale University Press, 2005.

Johannesburg Manifesto, Fairness in a Fragile World, Berlin: Heinrich Böll Foundation, 2002.

US Defence Dept, *An Abrupt Climate Change Scenario and It's Implications for US Natural Security*, 2003.

WWF, Living Planet Report, WWF International, 2006.

Further websites

The Edge
www.at-the-edge.org.uk

CABE
www.cabe.org.uk

China Dialogue
www.chinadialogue.net

Global Commons Institute (Contraction and Convergence)
www.gci.org.uk

Authors

Simon Foxell

Simon Foxell is the founding principal of The Architects Practice and a member of the Edge. He is a past chair of Policy and Strategy at the RIBA and currently acts as Design Advisor to Transforming Education, Birmingham City Council. He is the author of the *RIBA best practice guide to Starting a practice,* 2006, and *Mapping London: making sense of the city,* 2007.

Adam Poole

Adam Poole of Engineering Relations is also Reader at Ramboll Whitbybird. He previously ran an African-affairs consultancy whose projects included helping return Nigeria to democracy. He is a member of the edge..

Geoff Mulgan

Geoff Mulgan is the Director of the Young Foundation and was previously director of the government's Strategy Unit and head of policy in the Prime Minister's office. He was the founder and director of the think-tank Demos and is a board member of the Work Foundation and the Design Council. He is the author of several books including, *Good and Bad Power: the ideals and betrayals of government,* 2006.

The Edge

The Edge is a ginger group and think tank, sponsored by the building industry professions, that seeks to stimulate public interest in policy questions that affect the built environment, and to inform and influence public opinion. It was established in 1996 with support from the Arup Foundation. The Edge is supported by The Carbon Trust.

The Edge organises a regular series of debates and other events intended to advance policy thinking in the built environment sector and among the professional bodies within it. For further details, see www.at-the-edge.org.uk

Edge Futures

Edge Futures is a project initiated by The Edge and Black Dog Publishing. It has only been possible with the active participation of The Edge Committee as well as supporting firms and institutions. Special thanks are due to Adam Poole, Duncan McCorquodale, Frank Duffy, Robin Nicholson, Bill Gething, Chris Twinn, Andy Ford, Mike Murray and Jane Powell as well as to all the individual authors.

The project has been generously sponsored by The Carbon Trust, The Commission for Architecture and the Built Environment (CABE), Ramboll Whitbybird, The Arup Foundation, ProLogis and Construction Skills. Thanks are due to all those bodies and to the support of Karen Germain, Elanor Warwick, Mark Whitby, Ken Hall and Guy Hazlehurst within them.
The Edge is also grateful to Sebastian Macmillan of IDBE in Cambridge for the day we spent developing scenarios there and to Philip Guildford for facilitating the session.

Simon Foxell

Editorial committee for Edge Futures

- Frank Duffy
- Simon Foxell
- Duncan McCorquodale
- Adam Poole

The Edge is supported by:

- The Carbon Trust
- CIBSE
- ICE
- RIBA
- RICS
- IStructE

The Edge Committee

Chris Beauman	European Bank for Reconstruction and Development
Dr Bill Bordass	William Bordass Associates
Paddy Conaghan	Hoare Lea
Michael Dickson PPIStructE CBE	Buro Happold
Dr Frank Duffy PPRIBA CBE	DEGW
Dr Garry Felgate	
Rachel Fisher	
Andy Ford	Fulcrum Consulting
Prof Max Fordham PPCIBSE OBE	Max Fordham LLP
Simon Foxell	The Architects Practice
Prof Bill Gething	Feilden Clegg Bradley Studios
Jim Green	Baylight Properties
Prof Peter Guthrie	University of Cambridge
David Hampton	The Carbon Coach
Dr Jan Hellings	
Paul Hyett PPRIBA	RyderHKS
Prof Paul Jowitt	Heriot-Watt University
Janet Kidner	Lend Lease
Chani Leahong	Fulcrum Consulting
Duncan McCorquodale	Black Dog Publishing
Prof Mike Murray	Skanska
Robin Nicholson CBE	Edward Cullinan Architects
Michael Pawlyn	Exploration Architecture
Adam Poole	Engineering Relations
Andrew Ramsay	Engineering Council
Bruno Reddy	Arup
Yasmin Sharrif	Dennis Sharp Architects
Dr David Strong	Inbuilt Ltd
Chris Twinn	Arup
Bill Watts	Max Fordham LLP
Prof Mark Whitby PPICE	Ramboll Whitbybird
Terry Wyatt PPCIBSE	Hoare Lea & Partners

Much is already known about the state of the world 15 to 20 years from now. Almost all the buildings and infrastructure are already in place or in development—we replace our buildings etc., at a very slow pace. The great majority of the population who'll be living and working then, especially in the UK, have already been born and will have been educated in a school system that is familiar and predictable. The global population, however, will have increased from 6.7 billion in July 2007 to approximately 8 billion by 2025.

The climate will have changed, mainly as a result of the emissions of greenhouse gases of the past 50 and more years, but not by much. The temperature is predicted to be, on average, half a degree warmer, as well as varying over a greater range than at present. But, more significantly it will be understood to be changing, resulting in a strong feeling of uncertainty and insecurity. Rainfall will have reduced but will also become more extreme, i.e. tending to drought or flood. Resources, whether energy, water or food imports, will be in shorter supply; partly as a result of climate change but also due to regulations aimed at preventing the effects of global warming becoming worse. Transport will be constrained as a result but other technologies will have greatly improved the ability to economically communicate.

These changes form the context for this first series of five Edge Futures books, but it is not their subject: that is the impact of such changes and other developments on our daily lives, the economy, social and education services and the way the world trades and operates. Decision makers are already being challenged to act and formulate policy, in the face of the change already apparent in the years ahead. This set of books highlights how critical and important planning for the future is going to be. Society will expect and require policy makers to have thought ahead and prepared for the best as well as the worst. Edge

Futures offers a series of critical views of events, in the next two decades, that need to be planned for today.

The five books intentionally look at the future from very different viewpoints and perspectives. Each author, or pair of authors, has been asked to address a different sector of society, but there is inevitably a great deal of crossover between them. They do not always agree; but consistency is not the intention; that is to capture a breadth of vision as where we may be in 20 years time.

Jonathon Porritt in *Globalism and Regionalism* examines some of the greatest challenges before the planet, including climate change and demographic growth, and lays down the gauntlet to the authors of the other books. Porritt's diagnosis of the need to establish a new balance between the global and the regional over the years ahead and to achieve a 'Civic Globalisation' has an echo in Geoff Mulgan's call in *Living and Community* for strengthening communities through rethinking local governance and rebuilding a sense of place. Both are—perhaps professionally—optimistic that the climate change is a challenge that we, as a society, can deal with, while not underestimating the change that our society is going to have to undergo to achieve it.

Hank Dittmar, writing in *Transport and Networks* is less than certain, that currently, policies are adequately joined-up to deal with the issues that the recent flurry of major reports from the UK Government has highlighted: "Planning" from Barker, "Climate Change" from Stern and "Transport" from Eddington. He notes Barker's comment that "planning plays a role in the mitigation of and adaptation to climate change, the biggest issue faced across all climate areas"but that she then goes on to dismiss the issue. In its approach to all these reviews, the government has shown that it is more concerned with economic growth and indeed it has already concluded that the transport network needs no further fundamental reform. Dittmar believes otherwise, he calls for immediate solutions to support the development of the accessible, sustainable city.

Simon Foxell in *Education and Creativity* sees an even bumpier ride ahead, with progress only being made as a result of the lurch from crisis to crisis. Such discontinuities, will allow the UK to address many longstanding problems, from the personalisation of education to addressing the increasingly cut-throat international competition in creativity, innovation and skills—but not without a great deal of pain and chaos. Bill Mitchell, in the same volume, outlines a way of reconfiguring educational practice to develop just those skills that successful creativity-based economies are going to require.

In *Working*, Frank Duffy sees the end of road for the classic 'American Taylorist' office and the unsuitability of its counterpart, the European social democratic office. In their place, he proposes a new typology—the networked office—that will make better use of the precious resource that is our existing stock of buildings and allow greater integration into the life of the city. And, it is the city that all the authors come back to as a central and unifying theme—the dominant form of the millennium, the place where the majority of mankind now lives. Perhaps this is because, as Deyan Sudjic, Director of the Design Museum, has written recently; "The future of the city has suddenly become the only subject in town."

It is about the largest social unit that most of us can imagine with any ease and is a constant challenge economically, socially and environmentally. If we can work out what a sustainable city might be like and how to deliver it, then maybe we can sleep easier in our beds, less afraid that the end of civilisation, as we recognise it, may be within our childrens', or our childrens' childrens', lifetime. All the component parts of the Edge Futures studies come together in the city; where the community meets the office buildings, the schools and transport system. The city is the hub of the regional response to world events and needs to become a responsive participant in formulating a way out of policy log-jam.

As this first series of Edge Futures shows, the task is urgent and deeply complex but also not impossible. It is only, assuming that we need to make the transition to a low carbon economy within ten to twenty years, in Geoff Mulgan's words: "extraordinarily challenging by any historic precedent."

10a Acton Street
London WC1X 9NG
T. +44 (0)20 7613 1922
F. +44 (0)20 7613 1944
E. info@blackdogonline.com
W. www.blackdogonline.com

Designed by Draught Associates

British Library Cataloguing-in-Publication Data.
A CIP record for this book is available from the British Library.
ISBN: 978 1 906155 131

Black Dog Publishing, London, UK is an environmentally responsible company. Edge Futures are printed on Cyclus Offset, a paper produced from 100% post consumer waste.

architecture art design
fashion history photography
theory and things

www.blackdogonline.com

PATIENCE WORTH

A PSYCHIC MYSTERY

by

CASPER S. YOST

www.whitecrowbooks.com

PREFACE

THE editor of this book is not a spiritualist, nor a psychologist, nor a member of the Society for Psychical Research; nor has he ever had anything more than a transitory and skeptical interest in psychic phenomena of any character. He is a newspaper man whose privilege and pleasure it is to present the facts in relation to some phenomena which he does not attempt to classify nor to explain, but which are virtually without precedent in the record of occult manifestations. The mystery of Patience Worth is one which every reader may endeavor to solve for himself. The sole purpose of this narrative is to give the visible truth, the physical evidence, so to speak, the things that can be seen and that are therefore susceptible of proof by ocular demonstration. In this category are the instruments of communication and the communications themselves, which are described, explained and, in some cases, interpreted, where an effort at interpretation seems to be desirable.

Patience Worth: A Psychic Mystery

Published and printed in the United States of America and the United Kingdom
by White Crow Books; an imprint of White Crow Productions Ltd.

For information, contact White Crow Books
at P. O. Box 1013 Guildford, GU1 9EJ United Kingdom,
or e-mail to info@whitecrowbooks.com.

Cover Designed by Butterflyeffect
Interior production by essentialworks.co.uk
Interior design by Perseus Design

Paperback ISBN 978-1-908733-06-1
eBook ISBN 978-1-908733-07-8

Non Fiction / metaphysical / Parapsychology

Published by White Crow Books
www.whitecrowbooks.com

Contents

Chapter 1

THE COMING OF PATIENCE WORTH

ON a July evening in 1913 two women of St. Louis sat with a ouija board upon their knees. Some time before this a friend had aroused their interest in this unfathomable toy, and they had since whiled away many an hour with the inscrutable meanderings of the heart-shaped pointer; but, like thousands of others who had played with the instrument, they had found it, up to this date, but little more than a source of amused wonder. The messages which they had laboriously spelled out were only such as might have come from the sub-consciousness of either one or the other, or, at least, were no more strange than innumerable communications which have been received through the reading of the Ouija board.

But upon this night they received a visitor. The pointer suddenly became endowed with an unusual agility, and with great rapidity presented this introduction:

"Many moons ago I lived. Again I come. Patience Worth my name."

The women gazed, round-eyed, at each other, and the board continued:

"Wait. I would speak with thee. If thou shalt live, then so shall I. I make my bread by thy hearth. Good friends, let us be merrie. The time for work is past. Let the tabbie drowse and blink her wisdom to the firelog.

"How quaint that is!" one of the women exclaimed.

"Good Mother Wisdom is too harsh for thee," said the board, "and thou shouldst love her only as a foster mother."

Thus began an intimate association with "Patience Worth" that still continues, and a series of communications that in intellectual vigor and literary quality are virtually without precedent in the scant imaginative literature quoted in the chronicles of psychic phenomena.

The personality of Patience Worth—if personality it may be called—so impressed itself upon these women, at the first visit, that they got pencil and paper and put down not only all that she transmitted through the board, but all the questions and comment that elicited her remarks; and at every meeting since then, a verbatim record has been made of the conversation and the communications.

These records have accumulated until they have filled several volumes of typewritten pages, and upon them, and upon the writer's personal observations of the workings of the phenomena, this narrative is based. They include conversations, maxims, epigrams, allegories, tales, dramas, poems, all the way from sportive to religious, and even prayers, most of them of no little beauty and of a character that may reasonably be considered unique in literature.

The women referred to are Mrs. John H. Curran, wife of the former Immigration Commissioner of Missouri, and Mrs. Emily Grant Hutchings, wife of the Secretary of the Tower Grove Park Board in St. Louis, both ladies of culture and refinement. Mrs. Curran is a young woman of nervous temperament, bright, vivacious, and ready of speech. She has a taste for literature, but is not a writer, and has never attempted to write anything more ambitious than a personal letter. Mrs. Hutchings, on the other band, is a professional writer of skill, and it was to her quick appreciation of the quality of the communications that the starting of the record is due. It was soon apparent, however, that it was Mrs. Curran who was the sole agent of transmission; for the communications came only when she was at the board, and it mattered not who else sat with her. During the first months only Mrs. Curran and Mrs. Hutchings sat, but gradually the circle widened, and others assisted Mrs. Curran. Sometimes as many as five or six would sit with her in the course of an evening. Mr. Curran has acted as amanuensis, and recorded the communications at most of the sittings, Mrs. Curran's mother, Mrs. Mary E. Pollard, occasionally taking his place.

The Ouija board is a rectangular piece of wood about 16 inches wide by 24 inches in length and half an inch thick. Upon it the letters of the alphabet are arranged in two concentric arcs, with the ten numerals below, and the words "Yes" and "No" at the upper corners. The planchette, or pointer, is a thin, heart-shaped piece of wood provided with three legs, upon which it moves about upon the board, its point indicating the letters of the words it is spelling. Two persons are necessary for its operation. They place the tips of their fingers lightly upon the pointer and wait. Perhaps it moves; perhaps it does not. Sometimes it moves aimlessly about the board, spelling nothing; sometimes it spells words, but is unable to form a sentence; but often it responds readily enough to the impulses which control it, and even answers questions intelligibly, occasionally in a way that excites the wonder and even the awe of those about it. Its powers have been attributed by some to supernatural influence, by others to sub-consciousness, but science has looked upon it with disdain, as, until recent years, science has looked upon nearly all unprecedented phenomena.

Mr. W. T. Carrington, an eminent English investigator of psychical phenomena, in an exhaustive work upon the subject, has this to say of the Ouija board: "Granting for the sake of argument that the board is moved by the sitter, either consciously or unconsciously, the great and vital question still remains: What is the intelligence behind the board, that directs the phenomena? Whoever sets out to give a final and decisive answer to this question in the present state of our knowledge will have his task cut out for him, and I wish him happiness in the undertaking. Personally I am attempting nothing of the kind."

The Ouija board has been in use for many years. There is no element of novelty in the mere fact that curious and puzzling messages are received by means of it. I emphasize this fact because I wish to place the board in its proper relation to the communications from the intelligence calling herself Patience Worth. Aside from the psychical problem involved—and which, so far as the board is concerned, is the same in this case as in many others—the Ouija board has no more significance than a pen or a pencil in the hand. It is merely an instrument for the transmission of thought in words. In comparison with the personality and the literature which it reveals in this instance, it is a factor of little significance. It is proper to say, however, at this point, that every word attributed to Patience Worth in this volume was received by Mrs. Curran through this instrument.

Chapter 2

NATURE OF THE COMMUNICATIONS

"He who buildeth with peg and cudgel but buildeth a toy for an age who will but cast aside the bauble as naught; but he who buildeth with word, a quill and a fluid, buildeth well."—Patience Worth.

THERE are a number of things that distinguish Patience Worth from all other "intelligences" that have been credited with communications pretending to come from a spiritual source. First is her intellect. One of the strongest arguments against the genuineness of such communications has been the lack of intelligence often displayed in them. They have largely been, though with many exceptions, crude emanations of weak mentalities, and few of the exceptions have shown greater intellect or greater knowledge than is possessed by the average human being.

In a work entitled, Is Death the End? Dr. John H. Holmes, an eminent New York divine, gives considerable space to the psychic evidence of immortality. In the course of his discussion of this phase of his subject he concisely describes the characteristic features of psychic communications. "Nobody," he says, "can study, the evidence gathered in this particular field without noticing, first of all, the triviality, almost the inanity, of the communications received. Here we come, eager for the evidence of future life and information as to what it means to die and pass into the great beyond. And what do we get? First of all—and naturally enough, perhaps—frantic efforts on the part of the alleged spirits to prove their identity by the citation of intricate and unimportant details of where they were and what they did at different times when they were here among men. Sometimes there is a recounting of an event which is taking place in a part of the world far removed from the locality in which the medium and the recipient are sitting. Again and again there is a descent to obscurity and feeble chattering."

I quote this passage, not merely because it so clearly states the experience and conclusions of many who have investigated these phenomena, but because it serves to show by its marked contrast the wonder of the communications from Patience Worth. There are no efforts on her part to prove her identity. On the contrary, she can rarely be induced to speak of herself, and the personal information she has reluctantly given is disappointingly meager. "About me," she says, "thou wouldst know much. Yesterday is dead. Let thy mind rest as to the past." She never speaks of her own acts as a physical being; she never refers to any event taking place in the world now or that has taken place in the past. But far more important than these, she reveals an intellect that is worthy of any man's respect. It is at once keen, swift, subtle and profound. There is not once but always a sustained level of clear thought and fine feeling." There is obscurity at times, but it is usually the obscurity of profundity, and intelligent study generally reveals a meaning that is worth the effort. There is never a "focusing of attention upon the affairs of this world," except for the purpose of displaying its beauties and its wonders, and to assist in explaining the world that she claims is to come. For that other world she seems to try to explain as far as some apparent limitations permit, speaks as few have spoken before, and her words often bring delight to the mind and consolation to the soul.

Before considering these communications in detail, it would be well for the reader to become a little better acquainted with the alleged Patience herself. I speak of her as a person, for whatever she, or it, may be, the impression of a distinct personality is clear and definite; and it is, besides, more convenient so to designate her. Patience as a rule speaks an archaic tongue that is in general the English language of about the time of the Stuarts, but which contains elements of a usage still more ancient, and, not rarely, word and phrase forms that seem never to have been used in English or in any English dialect. Almost all of her words, however, whether in conversation or in literary composition, are of pure Anglo-Saxon-Norman origin. There is seldom a word of direct Latin or Greek parentage. Virtually all of the objects she refers to are things that existed in the seventeenth century or earlier. In all of the great mass of manuscript that has come from her we have not noticed a single reference to an object of modern creation or development; nor have more than a dozen words been found in her writings that may be of later origin than the seventeenth century, and some of these words are debatable. She has shown, in what would seem to be a genuinely feminine spirit of perversity, that she can use a modern word if she chooses to do so. And if she is living now, no matter when she was on earth, why should she not? (She has twice used the word 'shack', meaning a roughly constructed cabin, a word which is in that sense so new and so local that it has but recently found a place in the dictionaries.) But the fact remains that the number of such words is so small as to be negligible.

Only one who has tried to write in archaic English without committing anachronisms can realize its tremendous difficulty. We are so saturated with words and idioms of modern origin that it is almost impossible wholly to discard them, even when given every advantage of time and reflection. How much more difficult must it be then to use and maintain such language without an error in ordinary impromptu conversation, answering questions that could not have been expected, and flashing repartee that is entirely dependent upon the situation or remarks of the moment. Yet Patience does this with marvelous facility. So she can hardly be Mrs. Curran.

All of her knowledge of material things seems to be drawn from English associations. She is surprisingly familiar with the trees and flowers, the birds and beasts of England. She knows the manners and customs of its people as they were two or three centuries ago, the people of the fields or the people of the palace. Her speech is filled with references to the furniture, utensils and mechanical contrivances of the household of that time, and to its articles of dress, musical instruments, and tools of agriculture and the mechanical arts. There are also a few indications of knowledge of New England life. Yet she has never admitted a residence in England or New England, has never spoken of a birthplace or an abiding place anywhere, has never, in fact, used a single geographical proper name in relation to herself.

The communications of Patience Worth come in a variety of forms: Conversation that is strewn with wit and wisdom, epigrams and maxims; poems by the hundred; parables and allegories; stories of a semi-dramatic character, and dramas.

Here is an example of her conversation from one of the early records—an evening when a skeptical friend, a young physician, somewhat disposed to the use of slang, was present with his wife.

As the ladies took the board, the doctor remarked:

"I hope Patience Worth will come. I'd like to find out what her game is."

Patience was there and instantly responded:

"Dost, then, desire the plucking of another goose?"

Doctor.—"By George, she's right there with the grease, isn't she?"

Patience.—"Enough to baste the last upon the spit."

Doctor.—"Well, that's quick wit for you. Pretty hard to catch her."

Patience.—"The salt of today will not serve to catch the bird of tomorrow."

Doctor.—"She'd better call herself the bird of yesterday. I wonder what kind of a mind she had, anyway."

Patience.—"Dost crave to taste the sauce?

Doctor.—"She holds to her simile of the goose. I wish you'd ask her how she makes that little table move under your hands to spell the words."

Patience.—"A wise cook telleth not the brew."

Doctor.—"Turn that board over and let me see what's under it."

This was done, and after his inspection it was reversed.

Patience.—"Thee'lt bump thy nose to look within the hopper."

Doctor.—"Whew! She doesn't mind handing you one, does she?"

Mrs. Pollard.—"That's Patience's way. She doesn't think we count for anything."

Patience.—"The bell-cow doth deem the good folk go to Sabboth house from the ringing of her bell."

Doctor.—"She evidently thinks we are a conceited lot. Well, I believe she'll agree with me that you can't get far in this world without a fair opinion of yourself."

Patience.—"So the donkey loveth his bray!"

The Doctor's Wife.—"You can draw her on all you please. I'm going to keep perfectly still."

Patience.—"Oh, e'en the mouse will have a nibble."

Mrs. Curran.—"There! She isn't going to let you off without a little roast. I wonder what she has to say to you."

Patience.—"Did'st ever see the brood hen puff up with self-esteem when all her chicks go for a swim?"

Doctor.—"Let's analyze that and see if there's anything in it."

Patience.—"Strain the potion. Mayhap thou wilt find a fly."

This will be sufficient to illustrate Patience's form of speech and her ready wit. It also shows something of the character of the people to whom and through whom she has usually spoken. They are not solemn investigators or "pussy-footed" charlatans. There is no ceremony about the sitting, no dimmed lights, no compelled silences, no mummeries of any sort.

The assistance is of the ordinary, fun-loving, somewhat irreverent American type. The board is brought into the living room under the full glare of the electric lamps. The men perhaps smoke their cigars. If Patience seems to be in the humor for conversation, all may take part, and she hurls her javelins impartially. A visitor is at once brought within the umbra of her wit.

Her conversation, as already indicated, is filled with epigrams and maxims. A book could be made from these alone. They are, of course, not always original. What maxims are? But they are given on the instant, without possibility of previous thought, and are always to the point. Here are a few of these prompt aphorisms:

"A lollypop is but a breeder of pain."

"An old goose gobbles the grain like a gosling." "Dead resolves are sorry fare."

"The goose knoweth where the bin leaketh." "Quills of sages were plucked from geese."

"Puddings fit for lords would sour the belly of the swine boy."

"To clap the cover on a steaming pot of herbs will but modify * the stench."

"She who quacketh loudest deems the gander not the lead at waddling time."

"Climb not the stars to find a pebble."

"He who hath a house, a hearth and a friend hath a lucky lot."

She is often caustic and incisive.

"A man loveth his wife, but, ah, the buckles on his knee breeks!"

"Should I present thee with a pumpkin, wouldst thou desire to count the seeds?"

"A drink of asses' milk would nurture the swine, but wouldst thou then expect his song to change from Want, Want, Want?"

"Some folk, like the bell without a clapper, go clanging on in good faith, believing the good folk can hear them."

"Were I to tell thee the pudding string were a spinet's string, thou wouldst make ready for the dance."

"Thee'lt tie thy God within thy kerchief, else have none of Him, and like unto a bat, hang thyself topsy-turvy to better view His handiwork."

"'Twould pleg thee sore should thy shadow wear cap and bells."

"From constant wishing the moon may tip for thee."

"Wouldst thou have a daisy blossom upon a thistle?"

"Ye who carry pigskins to the well and lace riot the hole are a tiresome lot."

"He who eateth a bannock well made flattereth himself should his belly not sour."

Aside from the dramatic compositions, some of which are of great length, most of the communications received from Patience have been in verse. There is rarely a rhyme, practically all being iambic blank verse in lines of irregular length. The rhythm is to come as soon as the hands are placed upon the planchette, and the evening is given over to the production of verse. At others, verses are mingled with repartee and epigram, but seldom is an evening spent without at least one poem coming. This was not the case in the earlier months, when many sittings were given up wholly to conversation. The poetry has gradually increased in volume, as if the earlier efforts of the influence had been tentative, while the responsiveness of the intermediary was being tested. So, too, the earlier verses were fragments.

A blighted bud may hold
A sweeter message than the loveliest flower.
For God hath kissed her wounded heart
And left a promise there.

A cloak of lies may clothe a golden truth.
The sunlight's warmth may fade its glossy black
To whitening green and prove the fault
Of weak and shoddy dye.

Oh, why let sorrow steel thy heart?
Thy busom is but its foster mother,
The world its cradle, and the loving home
Its grave.

Weave sorrow on the loom of love
And warp the loom with faith.

Such fragments, however, were but steps leading to larger things. A little later on this came:

So thou hast trod among the tansey tuft
And murr and thyme, and gathered all the garden's store,
And glutted on the lillie's sensuous sweet,
And let thy shade to mar the sunny path,
And only paused to strike the slender bumming bird,
Whose molten-tinted wing but spoke the song
Of fluttering joy, and in thy very hand
Turned to motley gray. Then thinkest thou
To build the garden back by trickery?

And then, some six months after her first visit, came the poem which follows, and which may be considered the real beginning of her larger works:

Long lines of leaden cloud; a purple sea;
White gulls skimming across the spray.

Oh dissonant cry! Art thou
The death cry of desire?
Ah, wail, ye winds,
And search ye for my dearest wish
Along the rugged coast, and down
Where purling waters whisper
To the rosy coral reef.
Ah, search! Ah, search!
And when ye return, bring ye the answering.

Do I stand and call unto the sea for answer?
Ah, wisdom, where art thou?
A gull but shows thee to the Southland,
And leaden sky but warneth thee of storm.
And wind, thou art but a changeling.
So, shall I call to thee? Not so.
I build not upon the spray,
And seek not within the smaller world,
For God dwelleth not abroad, but deep within.

There is spiritual significance, more or less profound, in nearly all of the poems. Some of the lines are obscure, but study reveals a meaning, and the more I, at least, study them, the more I have been impressed with the intellectual power behind them. It is this that makes these communications seem to stand alone among the numerous messages that are alleged to have come from "that undiscovered country."

An intense love of nature is expressed in most of the communications, whether in prose or verse, and also a wide knowledge of nature—not the knowledge of the scientist, but that of the poet.

All silver-laced with web and crystal-studded, hangs
A golden lily cup, as airy as a dancing sprite.
The moon hath caught a fleeting cloud, and rests in her embrace.
The bumblefly still hovers o'er the clover flower,
And mimics all the zephyr's song.
White butterflies,
Whose wings bespeak late wooing of the buttercup,
Wend home their way, the gold still clinging to their snowy gossamer.
E'en the toad, who old and moss-grown seems,
Is wabbled on a lilypad, and watches for the moon
To bid the cloud adieu and light him to his bunt
For fickle marsh flies who tease him through the day.
Why, every rose has loosed her petals,
And sends a pleading perfume to the moss
That creeps upon the maple's stalk, to tempt it hence
To bear a cooling draught. Round yonder trunk
The ivy clings and loves it into green.
The pansy dreams of coaxing goldenrod
To change her station, lest her modest flower
Be ever doomed to blossom 'neath the shadow of the wall.
And was not He who touched the pansy
With His regal robes and left their color there,
All wise to leave her modesty as her greatest charm?
Here snowdrops blossom 'neath a fringe of tuft,
And fatty grubs find rest amid the mold.
All love, and Love himself, is here,
For every garden is fashioned by his hand.
Are then the garden's treasures more of worth

Than ugly toad or mold?
Not so, for Love
May tint the zincy blue-gray murk
Of curdling fall to crimson, light-flashed summer tide.
Ah, why then question Love, I prithee, friend?

This is poetry, but there is something more than liquid sweetness in its lines. There is a truth. Deeper wisdom and a lore more profound and more mystical are revealed or delicately concealed in some of the others.

I searched among the bills to find His love,
And found but waving trees, and stones
Where lizards flaunt their green and slip to cool
A down the moss. I searched within the field
To find His treasure-trove, and found but tasseled stalk
And baby grain, encradled in a silky nest.
I searched deep in the rose's heart to find
His pledge to me, and steeped in honey, it was there.
Lo, while I wait, a vagabond with goss'mer wing
Hath stripped her of her loot and borne it all to me.
I searched along the shore to find His heart,
Ahope the lazy waves would bear it me;
And watched them creep to rest upon the sands,
Who sent them back again, asearch for me.
I sought amid a tempest for His strength,
And found it in its shrieking glee;
And saw man's paltry blocks come crashing down,
And heard the wailing of the trees who grew
A feared, and, moaning, caused the flowers to quake
And tremble lest the sun forget them at the dawn;
While bolts shot clouds asunder, and e'en the sea
Was panting with the spending of his might.
I searched within a wayside cot for His white soul,
And found a dimple next the lips of one who slept,
And watched the curtained wonder of her eyes,
Aflutter o'er the iris-colored pools that held His smile:
And touched the warm and shrinking lips, so mute,
And yet so wise.
For canst thou doubt whose kiss

Still lingers on their bloom?
Amid a muck of curse, and lie,
And sensuous lust, and damning leers,
I searched for Good and Light,
And found it there, aye, even there;
For broken reeds may house a lark's pure nest.
I stopped me at a pool to rest,
And toyed along the brink to pluck
The cress who would so guard her lips:
And flung a stone straight to her heart,
And, lo, but silver laughter mocketh me!
And as I stoop to catch the plash,
Pale sunbeams pierce the bower,
And ah, the shade and laughter melt
And leave me, empty, there.
But wait! I search and find,
Reflected in the pool, myself, the searcher.
And, on the silver surface traced,
My answer to it all.
For, heart of mine, who on this journey
Sought with me, I knew thee not,
But searched for prayer and love amid the rocks
Whilst thou but now declare thyself to me.
Ah, could I deem thee strong and fitting
As the tempest to depict His strength;
Or yet as gentle as the smile of baby lips,
Or sweet as honeyed rose or pure as mountain pool?
And yet thou art, and thou art mine
A gift and answer from my God.

It is not my purpose to attempt an extended interpretation of the metaphysics of these poems. This one will repay real study. No doubt there will be varied views of its meaning.

These poems do not all move with the murmuring ripple of running brooks. Some of them, appalling in the rugged strength of their figures of speech, are like the storm waves smashing their sides against the cliffs. In my opinion there are not very many in literature that grip the mind with greater force than the first two lines of the brief one which follows, and there are few things more beautiful than its conclusion:

Ah, God, I have drunk unto the dregs,
And flung the cup at Thee!
The dust of crumbled righteousness
Hath dried and soaked unto itself
E'en the drop I spilled to Bacchus,
Whilst Thou, all patient,
Sendest purple vintage for a later harvest.

The poems sometimes contain irony, gentle as a summer zephyr or crushing as a mailed fist. For instance this challenge to the vainglorious:

Strike ye the sword or dip ye in an inken well;
Smear ye a gaudy color or daub ye the clay?
Aye, beat upon thy busom then and cry,
'Tis mine, this world-love and vainglory!"
Ah, master-band, who guided thee? Stay!
Dost know that through the ages,
Yea, through the very ages,
One grain of hero dust, blown from afar,
Hath lodged, and moveth thee?
Wait. Wreathe thyself and wait.
The green shall deepen to an ashen brown
And crumble then and fall into thy sightless eyes,
While thy moldering flesh droppeth awry.
Wait, and catch thy dust.
Mayhap thou canst build it back!

She touches all the strings of human emotion, and frequently thrums the note of sorrow, usually, however, as an overture to a paean of joy. The somber tones in her pictures, to use another metaphor, are used mainly to strengthen the high lights. But now and then there comes a verse of sadness such as this one, which yet is not wholly sad:

Ah, wake me not!
For should my dreaming work a spell to soothe
My troubled soul, wouldst thou deny me dreams?
Ah, wake me not!
If 'mong the leaves wherein the shadows lurk
I fancy conjured faces of my loved, long lost;
And if the clouds to me are sorrow's shroud;

And if I trick my sorrow, then, to bide
Beneath a smile; or build of wasted words
A key to wisdom's door—wouldst thou deny me?
Ah, let me dream!
The day may bring fresh sorrows,
But the night will bring new dreams.

When this was spelled upon the board, its pathos affected Mrs. Curran to tears, and, to comfort her, Patience quickly applied an antidote in the following jingle, which illustrates not only her versatility, but her sense of humor:

Patter, patter, briney drops,
On my kerchief drying:
Spatter, spatter, salty stream,
Down my poor cheeks flying.
Brine enough to 'merse a ham,
Salt enough to build a dam!
Trickle, trickle, all ye can
And wet my dry heart's aching.
Sop and sop, 'tis better so,
For in dry soil flowers ne'er grow.

This little jingle answered its purpose. Mrs. Curran's tears continued to fall, but they were tears of laughter, and all of the little party about the board were put in good spirits. Then Patience dryly remarked:

"Two singers there be; he who should sing like a troubadour and brayeth like an ass, and he who should bray that singeth."

These examples will serve to illustrate the nature of the communications, and as an introduction to the numerous compositions that will be presented in the course of this narrative.

The question now arises, or, more likely, it has been in the reader's mind since the book was opened: What evidence is there of their genuineness?

Does Mrs. Curran, consciously or subconsciously, produce this matter? It is hardly credible that anyone able to write such poems would bother with a Ouija board to do it.

It will probably be quite evident to a reader of the whole matter that whoever or whatever it is that writes this poetry and prose, possesses, as already intimated, not only an unusual mind, but an unusual knowledge of archaic forms of English, a close acquaintance with nature as it is found in England, and a familiarity with the manners and customs of English life of an older time. Many of the words used in the later compositions, particularly those of a dramatic nature, are obscure dialectal forms not to be found in any work of literature. All of the birds and flowers and trees referred to in the communications are native to England, with the few exceptions that indicate some knowledge of New England. No one not growing up with the language used could have acquired facility in it without years of patient study. No one could become so familiar with English nature without long residence in England: for the knowledge revealed is not of the character that can be obtained from books. Mrs. Curran has had none of these experiences. She has never been in England. Her studies since leaving school have been confined to music, to which art she is passionately attached, and in which she is adept. She has never been a student of literature, ancient or modern, and has never attempted any form of literary work. She has bad no particular interest in English history, English literature or English life.

But, it may be urged, this matter might be produced subconsciously, from Mrs. Curran's mind or from the mind of some person associated with her. The phenomena of sub-consciousness are many and varied, and the word is used to indicate, but does not explain, numerous mysteries of the mind which seem wholly baffling despite this verbal hitching post. But I have no desire to enter into an argument. My sole purpose is so to present the facts that the reader may intelligently form his own opinion. Here are the facts that relate to this phase of the subject:

Mrs. Curran does not go into a trance when the communications are received. On the contrary, her mind is absolutely normal, and she may talk to others while the board is in operation under her hands. It is unaffected by conversation in the room. There is no effort at mental concentration. Aside from Mrs. Curran, it does not matter who is present, or who sits at the board with her; there are seldom the same persons at any two successive sittings. Yet the personality of Patience is constant and unvarying. As to subconscious action on the part of Mrs. Curran, it would seem to be sufficient to say that no one can impart knowledge subconsciously, unless it has been first acquired through the media of consciousness; that is to say, through the senses. No one, for example, who had never seen or heard a word of Chinese, could speak the language subconsciously. One may unconsciously acquire information, but it must be through the senses.

It remains but to add that the reputation and social position of the Currans puts them above the suspicion of fraud, if fraud were at all possible in such a matter as this; that Mrs. Curran does not give public exhibitions, nor private exhibitions for pay; that the compositions have been received in the presence of their friends, or of friends of their friends, all specially invited guests. There seems nothing abnormal about her. She is an intelligent, conscientious woman, a member of the Episcopalian church, but not especially zealous in affairs of religion, a talented musician, a clever and witty conversationalist, and a charming hostess. These facts are stated not as gratuitous compliments, but as evidences of character and temperament which have a bearing upon the question.

Chapter 3

PERSONALITY OF PATIENCE

Yea, I be me.

PATIENCE, as I have said, has given very little information about herself, and every effort to pin her to a definite time or locality has been without avail. When she first introduced herself to Mrs. Curran, she was asked where she came from, and she replied, "Across the sea." Asked when she lived, the pointer groped among the figures as if struggling with memory, and finally, with much hesitation upon each digit, gave the date 1649. This seemed to be so in accord with her language, and the articles of dress and household use to which she referred, that it was accepted as a date that had some relation to her material existence. But Patience has since made it quite plain that she is not to be tied to any period.

"I be like to the wind," she says, "and yea, like to it do blow me ever, yea, since time. Do ye to tether me unto today I blow me then tomorrow, and do ye to tether me unto tomorrow I blow me then today."

Indeed, she at times seems to take a mischievous delight in baffling the seeker after personal information; and at other times, when she has a composition in hand, she expresses sharp displeasure at such inquiries. As this is not a speculative work, but a narrative, the attempt to fix a time and place for her will be left to those who may find interest in the task. All that can be said with definiteness is that she brings the speech and the atmosphere, as it were, of an age or ages long past; that she is thoroughly English, and that while she can and does project herself back into the mists of time, and speak of early medieval scenes as familiarly as of the English renaissance, she does not make use of any knowledge she may possess of modern developments or modern conditions. And yet, archaic in word and form as her compositions are, there is something very modern in her way of thought and in her attitude toward nature. An eminent philologist asked her how it was that she used the language of so many different periods, and she replied: "I do plod a twist of a path and it, hath run from then till now." And when he said that in her poetry there seemed to be echoes or intangible suggestions of comparatively recent poets, and asked her to explain, she said: "There be aneath the every stone a hidden voice. I but loose the stone and lo, the voice!"

But while the archaic form of her speech and writings is an evidence of her genuineness, and she so considers it, she does not approve of its analysis as a philological amusement. "I brew and fashion feasts," she says, "and lo, do ye to tear asunder, thee wouldst have but grain dust and unfit to eat. I put not meaning to the tale, but source thereof." That is to say, she does not wish to be measured by the form of her words, but by the thoughts they convey and the source from which they come. And she has put this admonition into strong and striking phrases.

"Put ye a value 'pon word? And weigh ye the line to measure, then, the gift o' Him 'pon rod afashioned out by man?

"I tell thee, He hath spoke from out the lowliest, and man did put to measure, and lo, the lips astop!

"And He doth speak anew; yea, and He hath spoke from out the mighty, and man doth whine o' track ashow 'pon path he knoweth not—and lo, the mighty be astopped!

"Yea, and He ashoweth wonders, and man findeth him a rule, and lo, the wonder shrinketh, and but the rule remaineth!

"Yea, the days do rock with word o' Him, and man doth look but to the rod, and lo, the word o' Him asinketh to a whispering, to die.

"And yet, in patience, He seeketh new days to speak to thee. And thou ne'er shalt see His working. Nay!

"Look ye unto the seed o' the olive tree, aplanted. Doth the master, at its first burst athrough the sod, set up a rule and murmur him, "Tis ne'er an olive tree! It hath but a pulp stem and winged leaves?' Nay, he letteth it to grow, and nurtureth it thro' days, and lo, at finish, there astandeth the olive tree!

"Ye'd uproot the very seed in quest o' root! I bid thee nurture o' its day astead.

"I tell thee more: He speaketh not by line or word; Nay, by love and giving.

"Do ye also this, in His name."

But, aside from the meagerness of her history, there is no indefiniteness in her personality, and this clear-cut and unmistakable individuality, quite different from that of Mrs. Curran, is as strong an evidence of her genuineness as is the uniqueness of her literary productions. To speak of something which cannot be seen nor heard nor felt as a personality, would seem to be a misuse of the word, and yet personality is much more a matter of mental than of physical characteristics. The tongue and the eyes are merely instruments by means of which personality is revealed. The personality of Patience Worth is manifested through the instrumentality of a Ouija board, and her striking individuality is thereby as vividly expressed as if she were present in the flesh. Indeed, it requires no effort of the imagination to visualize her. Whatever she may be, she is at hand. Nor does she have to be solicited. The moment the fingers are on the board she takes command. She seems fairly to jump at the opportunity to express herself.

And she is essentially feminine. There are indubitable evidences of feminine tastes, emotions, habits of thought, and knowledge. She is, for example, profoundly versed in the methods of housekeeping of two centuries or more ago. She is familiar with all the domestic machinery and utensils of that olden time—the operation of the loom and the spinning wheel, the art of cooking at an open hearth, the sanding of floors; and this homely knowledge is the essence of many of her proverbs and epigrams.

"A good wife," she says, "keepeth the floor well sanded and rushes in plenty to burn. The pewter should reflect the fire's bright glow; but in thy day housewifery is a sorry trade."

At another time she opened the evening thus:

"I have brought me some barley corn and a porridge pot. May I then sup?"

And the same evening she said to Mrs. Pollard:

"Thee'lt ever stuff the pot and wash the dishcloth in thine own way. Alackaday! Go brush thy hearth. Set pot aboiling. Thee'lt cook into the brew a stuff that tasteth full well unto thy guest."

A collection of maxims for housekeepers might be made from the flashes of Patience's conversation. For example:

"Too much sweet may spoil the short bread."

"Weak yarn is not worth the knitting."

"A pound for pound loaf was never known to fail."

"A basting but toughens an old goose."

These and many others like them were used by her in a figurative sense, but they reveal an intimate knowledge of the household arts and appliances of a forgotten time. If she knows anything of stoves or ranges, of fireless cookers, of refrigerators, of any of the thousand and one utensils which are familiar to the modern housewife, she has never once let slip a word to betray such knowledge.

At one time, after she had delivered a poem, the circle fell into a discussion of its meaning, and after a bit Patience declared they were "like treacle dripping," and added, "thee'lt find the dishcloth may make a savory stew."

"She's roasting us," cried Mrs. Hutchings.

"Nay," said Patience, "boiling the pot."

"You don't understand our slang, Patience," Mrs. Hutchings explained. "Roasting means criticising or rebuking."

"Yea, basting," said Patience.

Mrs. Pollard remarked: "I've heard my mother say, 'He got a basting!'"

"An up-and-down turn to the hourglass does to a turn," Patience observed dryly.

"I suppose she means," said Mrs. Hutchings, "that two hours of basting or roasting would make us understand."

"Would she be likely to know about hourglasses?" Mrs. Curran asked.

Patience answered the question.

"A dial beam on a sorry day would make a muck o' basting." Meaning that a sundial was of no use on a cloudy day.

But Patience is not usually as patient with lack of understanding as this bit of conversation would indicate.

"I dress and baste thy fowl," she said once, and thee wouldst have me eat for thee. If thou wouldst build the comb, then search thee for the honey."

"Oh, we know we are stupid," said one. "We admit it."

"Saw drip would build thy head and fill thy crannies", Patience went on, "yet ye feel smug in wisdom."

And again: "I card and weave, and ye look a painful lot should I pass ye a bobbin to wind."

A request to repeat a doubtful line drew forth this exclamation: "Bother! I fain would sew thy seam, not do thy patching."

At another time she protested against a discussion that interrupted the delivery of a poem: "Who then doth hold the distaff from whence the thread doth wind? Thou art shuttling 'twixt the woof and warp but to mar the weaving."

And once she exclaimed, "I sneeze on rust o' wits!"

But it must not be understood that Patience is bad- tempered. These outbreaks are quoted to show one side of her personality, and they usually indicate impatience rather than anger: for, a moment after such caustic exclamations, she is likely to be talking quite genially or dictating the tenderest of poetry. She quite often, too, expresses affection for the family with which she has associated herself. At one time she said to Mrs. Curran, who had expressed impatience at some cryptic utterance of the board:

Ah, weary, weary me, from trudging and tracking o'er the long road to thy heart! Wilt thou, then, not let me rest awhile therein?"

And again: "Should thee let thy fire to ember I fain would cast fresh faggots."

And at another time she said of Mrs. Curran: "She doth boil and seethe, and brew and taste., but I have a loving for the wench."

But she seems to think that those with whom she is associated should take her love for granted, as home folks usually do, and she showers her most beautiful compliments upon the casual visitor who happens to win her favor. To one such she said:

"The heart o' her hath suffered thorn, but bloomed a garland o'er the wounds."

To a lady who is somewhat deaf she paid this charming tribute:

"She hath an ear upon her every finger's tip, and 'pon her eye a thousand flecks o' color for to spread upon a dreary tale and paint a leaden sky aflash. What need she o' ears?"

And to another who, after a time at the board, said she did not want to weary Patience:

"Weary then at loving of a friend? Would I then had the garlanded bloom o' love she hath woven and lighted, I do swear, with smiling washed brighter with her tears."

And again: "I be weaving of a garland. Do leave me then a bit to tie its ends. I plucked but buds, and woe! they did spell but infant's love. I cast ye, then, a blown bloom, wide petaled and rich o' scent. Take thou and press atween thy heart throbs—my gift."

Of still another she said: "She be a starbloom blue that nestleth to the soft grasses of the spring, but ah, the brightness cast to him who seeketh field aweary!"

And yet again: "Fields hath she trod arugged, aye, and weed agrown. Aye, and e'en now, where she hath set abloom the blossoms o' her very soul, weed aspringeth. And lo, she standeth head ahigh and eye to sky and faith astrong. And foot abruised still troddeth rugged field. But I do promise ye 'tis such an faith that layeth low the weed and putteth 'pon the rugged path asmoothe, and yet but bloom shalt show, and ever shalt she stand, head ahigh and eye unto the sky."

Upon an evening after she had showered such compliments upon the ladies present she exclaimed:

"I be a wag atruth, and lo, my posey-wreath be stripped!"

She seldom favors the men in this way. She has referred to herself several times as a spinster, and this may account for a certain reluctance to saying complimentary things of the other sex. "A prosy spinster may but plash in love's pool," she remarked once, and at another time she said: "A wife shall brush her goodman's blacks and polish o' his buckles, but a maid may not dare e'en to blow the trifling dust from his knickerbockers." With a few notable exceptions, her attitude toward men has been expressed in sarcasm, none the less cutting to those for whom she has an affection manifested in other ways. To one such she said:

"Thee'lt peg thy shoes, lad, to best their wearing, and eat too freely of the fowl. Thy belly needeth pegging sore, I wot."

"Patience doesn't mean that for me," he protested.

"Nay," she said, "the jackass ne'er can know his reflection in the pool. He deemeth the thrush hath stolen of his song. Buy thee a pushcart. 'Twill speak for thee."

And of this same rotund friend she remarked, when be laughed at something she had said:

"He shaketh like a pot o' goose jell!

"I back up, Patience," he cried.

"And thee'lt find the cart," she said.

Of a visitor, a physician, she had this to say:

"He bindeth and asmears and looketh at a merry, and his eye doth lie. How doth he smite and stitch like to a wench, and brew o'er steam! Yea, 'tis atwist he be. He runneth whither, and, at a beconing, (beckoning) yon, and ever thus; but 'tis a blunder-mucker he be. His head like to a steel, yea, and heart a summer's cloud athin (within), enough to show athrough the clear o' blue."

But it is upon the infant that Patience bestows her tenderest words. Her love of childhood is shown in many lines of rare and touching beauty.

"Ye seek to level unto her," she said of a baby girl who was present one evening, "but thou art awry at reasoning. For he who putteth him to babe's path doth track him high, and lo, the path leadeth unto the Door. Yea, and doth she knock, it doth ope.

"Cast ye wide thy soul's doors and set within such love. For, brother, I do tell thee that though the soul o' ye be torn, aye, and scarred, 'tis such an love that doth heal. The love o' babe be the balm o' earth.

"See ye! The sun tarrieth 'bout the lips o' her; aye, and though the hand be but thy finger's span, 'tis o' a weight to tear away thy heart."

And upon another occasion she revealed something of herself in these words:

Know ye; in my heart's mansion
There be apart a place
Wherein I treasure my God's gifts.
Think ye to peer therein?
Nay.
And should thee by a chance
To catch a stolen glimpse,
Thee'dst laugh amerry, for hord (hoard)
Would show but dross to thee:
A friend's regard, ashrunked and turned
To naught—but one bright memory is there;
A hope—now dead, but sboweth gold bid there;
A host o' nothings—dreams, hopes, fears;
Love throbs afluttered hence
Since first touch o' baby bands
Caressed my heart's store abidden.

Returning to the femininity of Patience, it is also shown in her frequent references to dress. Upon an evening when the publication of her poems bad been under discussion, when next the board was taken up she let them know that she had heard, in this manner:

"My pettieskirt hath a scallop," she said. 'Mayhap that will help thy history."

"Oh," cried Mrs. Curran, "we are discovered!"

"Yea," laughed Patience—she must have laughed, "and tell thou of my buckled boots and add a cap-string."

Further illustrative of her feminine characteristics and of her interest in dress, as well as of a certain fun-loving spirit which now and then seems to sway her, is this record of a sitting upon an evening when Mr. Curran and Mr. Hutchings had gone to the theater, and the ladies were alone:

Patience.—"Go ye to the lighted hall to search for learning? Nay, 'tis a piddle, not a stream, ye search. Mayhap thou sendest thy men for barleycorn. 'Twould then surprise thee should the asses eat it."

Mrs. H.—"What is she driving at?

Mrs. P.—"The men and the theater, I suppose."

Mrs. H.—"Patience, what are they seeing up there?"

Patience.—"Ne'er a timid wench, I vum."

Mrs. C.—"You don't approve of their going, do you, Patience?"

Patience.—"Thee'lt find a hearth more profit.
Better they cast the bit of paper."

Mrs. C.—"What does she mean by paper? Their programmes?"

Patience.—"Painted parchment squares."

Mrs. P.—"Oh, she means they'd better stay at home and
play cards."

Mrs. H.—"Are they likely to get their morals corrupted at that show?"

Patience.—"He who tickleth the ass to start a braying, fain would carol with his brother."

Mrs. C.—"If the singing is as bad as it usually is at that place, I don't wonder at her usual disapproval. But what about the girls, Patience?"

Patience.—"My pettieskirt ye may borrow for the brazens."

Mrs. P.—"'Now, what is a pettieskirt? Is it really a skirt or is it that ruff they used to wear around the neck?"

Patience.—"Nay, my bib covereth the neckband." Mrs. H.—"Then, where do you wear your pettieskirt?"

Patience.—"'Neath my kirtle."

Mrs. C.—"Is that the same as girdle? Let's look it up."

Patience.—"Art fashioning thy new frock?"

Mrs. H.—"I predict that Patience will found a new style— Puritan."

Patience.—"'Twere a virtue, egad!

Mrs. H.—"You evidently don't think much of our present style. In your day women dressed more modestly, didn't they?"

Patience.—"Many's the wench who pulled her points to pop. But ah, the locks were combed to satin! He who bent above might see himself reflected."

Mrs. C.—"What were the young girls of your day like, Patience?"

Patience.—"A silly lot, as these of thine. Wait!"

There was no movement of the board for about three minutes, and then:

"'Tis a sorry lot, not harming but boresome.

Mrs. H.—"Oh, Patience, have you been to the theater?"

Patience.—"A peep in good cause could surely ne'er harm the godly."

Mrs. C.—"How do you think we ought to look after those men?"

Patience.—"Thine ale is drunk at the hearth. Surely he who stops to sip may bless the firelog belonging to thee."

When the men returned home they agreed with the verdict of Patience before they had heard it, that it was a "tame" show, "not harming, but boresome."

The exclamation of Mrs. Curran, "Let's look it up," in the extract just quoted from the record, has been a frequent one in this circle since Patience came. So many of her words are obsolete that her friends are often compelled to search through the dictionaries and glossaries for their meaning. Her reference to articles of dress—wimple, kirtle, pettieskirt, points and so on, had all to be "looked up." Once Patience began an evening with this remark:

The cockshut finds ye still peering to find the other land."

"What is cock's hut?" asked Mrs. H.

"Nay," said Patience, "Cock-shut. Thee needeth light, but cockshut bringeth dark."

"Cockshut must mean shutting up the cock at night," suggested a visitor.

"Aye, and geese, too, then could be put to quiet," Patience exclaimed. "Wouldst thou wish for cockshut?"

Search revealed that cockshut was a term anciently applied to a net used for catching woodcock, and it was spread at nightfall, hence cockshut acquired also the meaning of early evening. Shakespeare uses the term once, in Richard III., in the phrase, "Much about cockshut time," but it is a very rare word in literature, and probably has not been used, even colloquially, for centuries.

There are many such words used by Patience—relics of an age long past. The writer was present at a sitting when part of a romantic story-play of medieval days was being received on the board. One of the characters in the story spoke of herself as "playing the jane-o'-apes." No one present had ever heard or seen the word. Patience was asked if it had been correctly received, and she repeated it. Upon investigation it was found that it is a feminine form of the familiar jackanapes, meaning a silly girl. Massinger used it in one of his plays in the seventeenth century, but that appears to be the only instance of its use in literature.

These words may be not unknown to many people, but the point is that they were totally, strange to those at the board, including Mrs. Curran—words that could not possibly have come out of the consciousness or subconsciousness of any one of them. The frequent use of such words helps to give verity to the archaic tongue in which she expresses her thoughts, and the consistent and unerring use of this obsolete form of speech is, next to the character of her literary production, the strongest evidence of her genuineness. It will be noticed, too, that the language she uses in conversation is quite different from that in her literary compositions, although there are definite similarities which seem to prove that they come from the same source. In this also she is wholly consistent: for it is unquestionably true that no poet ever talked as he wrote. Every writer uses colloquial words and idioms in conversation that he would never employ in literature. No matter what his skill or genius as a writer may be, he talks "just like other people." Patience Worth in this, as in other things, is true to her character.

It may be repeated that in all this matter—and it is but a skimming of the mass—one may readily discern a distinct and striking personality; not a wraith-like, formless, evanescent shadow, but a personality that can be clearly visualized. One can easily imagine Patience Worth to be a woman of the Puritan period, with, however, none of the severe and gloomy beliefs of the Puritan—a woman of a past age stepped out of an old picture and leaving behind her the material artificialities of paint and canvas. From her speech and her writings one may conceive her to be a woman of Northern England, possibly: for she uses a number of ancient words that are found to have been peculiar to the Scottish border; a country woman, perhaps, for in all of her communications there are only two or three references to the city, although her knowledge and love of the drama may be a point against this assumption; a woman who had read much in an age when books were scarce, and women who could read rarer still: for although she frequently expresses disdain of book learning, she betrays a large accumulation of such learning, and a copious vocabulary, as well as a degree of skill in its use, that could only have been acquired from much study of books. "I have bought beads from a pack," she says, "but ne'er yet have I found a peddler of words."

And then, after we have mentally materialized this woman, and given her a habitation and a time, Patience speaks again, and all has vanished. Not so," she said to one who questioned her, I be abirthed awhither and abide me where." And again she likened herself to the wind. "I be like the wind," she said, "who leaveth not track, but ever 'bout, and yet like to the rain who groweth grain for thee to reap." At other times she has indicated that she has never had a physical existence. I have quoted her saying: "I do plod a twist o' a path and it hath run from then till now." At a later time she was asked what she meant by that. She answered:

"Didst e'er to crack a stone, and lo, a worm aharded? (a fossil). 'Tis so, for list ye, I speak like ye since time began."

It is thus she reveals herself clearly to the mind, but when one attempts to approach too closely, to lay a hand upon her, as it were, she invariably recedes into the unfathomable deeps of mysticism.

Chapter 4

THE POETRY

Am I a broken lyre,
Who, at the Master's touch,
Respondeth with a tinkle and a whir?
Or am I strung in full
And at His touch give forth the full chord?
—Patience Worth

As the reader will have observed, the poetry of Patience Worth is not confined to a single theme, or to a group of related themes. It covers a range that extends from inanimate things through all the gradations of material life and on into the life of spiritual realms as yet uncharted. It includes poems of sentiment, poems of nature, poems of humanity; but the larger number deal with man in relation to the mysteries of the beyond. All of them evince intellectual power, knowledge of nature and human nature, and skill in construction. With the exception of one or two little jingles, the poems are rhymeless. Patience may not wholly agree with Milton that rhyme "is the invention of a barbarous age to set off wretched matter and lame metre," but she seldom uses it, finding in blank verse a medium that suits all her moods, making it at will as light and ethereal as a summer cloud or as solemn and stately as a Wagnerian march. She molds it to every purpose, and puts it to new and strange uses. Who, for example, ever saw a lullaby in blank verse? It is, I believe, quite without precedent in literature, and yet it would not be easy to find a lullaby more daintily beautiful than the one which will be presented later on.

In all of her verse, the iambic measure is dominant, but it is not maintained with monotonous regularity. She appreciates the value of an occasional break in the rhythm, and she understands the uses of the pause. But she declines to be bound by any rules of line measurement. Many of her lines are in accord with the decasyllabic standard of heroic verse, but in no instance is that standard rigidly adhered to: some of the lines contain as many as sixteen syllables, others drop to eight or even six.

It should be explained, however, that the poetry as it comes from the Ouija board is not in verse form. There is nothing in the dictation to indicate where a line should begin or where end, nor, of course, is there any punctuation, there being no way by which the marks of punctuation could be denoted. There is usually, however, a perceptible pause at the end of a sentence. The words are taken down as they are spelled on the board, without any attempt, at the time, at versification or punctuation. After the sitting, the matter is punctuated and lined as nearly in accord with the principles of blank verse construction as the abilities of the editor will permit. It is not claimed that the line arrangement of the verses as they are here presented is perfect; but that is a detail of minor importance, and for whatever technical imperfections there may be in this particular, Patience Worth is not responsible. The important thing is that every word is given exactly as it came from the board, without the alteration of a syllable, and without changing the position or even the spelling of a single one.

As a rule, Patience spells the words in accordance with the standards of today, but there are frequent departures from those standards, and many times she has spelled a word two or three different ways in the same composition. For example, she will spell "spin" with one n or two n's indifferently: she will spell "friend" correctly, and a little later will add an e to it; she will write "boughs" and "bows" in the same composition. On the other hand she invariably spells tongue "tung," and positively refuses to change it, and this is true also of the word bosom, which she spells "busom."

There are indications that the poems and the stories are in course of composition at the time they are being produced on the Ouija board. Indeed, one can almost imagine the author dictating to an amanuensis in the manner that was necessary before stenography was invented, when every word had to be spelled out in long-hand.

At times the little table will move with such rapidity that it is very difficult to follow its point with the eye and catch the letter indicated.

Then there will be a pause, and the pointer will circle around the board, as if the composer were trying to decide upon a word or a phrase. Occasionally four or five words of a sentence will be given, then suddenly the planchette will dart up to the word "No," and begin the sentence again with different and, it is to be presumed, more satisfactory words.

Sometimes, though rarely, Patience will begin a composition and suddenly abandon it with an exclamation of displeasure, or else take up a new and entirely different subject. Once she began a prose composition thus:

"I waste my substance on the weaving of web and the storing of pebbles. When shall I build mine house, and when fill the purse? Oh, that my fancy weaved not but web, and desire pricketh not but pebble!"

There was an impatient dash across the board, and then she exclaimed:

"Bah, 'tis bally reasoning! I plucked a gosling for a goose, and found down enough to pad the parson's saddle skirts!

At another time she began:

"Rain, art thou the tears wept a thousand years agone, and soaked into the granite walls of dumb and feelingless races? Now— "There was a long pause and then came this lullaby:

Oh, baby, soft upon my breast press thou,
And let my fluttering throat spell song to thee,
A song that floweth so, my sleeping dear:
Oh, buttercups of eve,
Oh, willynilly,
My song shall flutter on,
Oh, willynilly.
I climb a web to reach a star,
And stub my toe against a moonbeam
Stretched to bar my way,
Oh, willynilly.
A love—puff vine shall shelter us,
Oh, baby mine;
And then across the sky we'll float

And puff the stars away.
Oh, willynilly, on we'll go,
Willynilly floating.

Thee art o'erfed on pudding," she added to Mrs. Curran. "This sauce is but a butterwhip."

And now, having briefly referred to the technique of the poems, and explained the manner in which they are transmitted we will make a more systematic presentation of them. For a beginning, nothing better could be offered than the Spinning Wheel lullaby heretofore referred to.

In it we can see the mother of, perhaps, the Puritan days, seated at the spinning wheel while she sings to the child which is supposed to lie in the cradle by her side. One can view through the open door the old-fashioned flower garden bathed in sunlight, can hear the song of the bird and the hum of the bee, and through it all the sound of the wheel. But!—it is the song of a childless woman to an imaginary babe: Patience has declared herself a spinster.

Strumm, strumm!

Ah, wee one,
Croon unto the tendrill tipped with sungilt,
Nodding thee from o'er the doorsill there.
Strumm, strumm!

My wheel shall sing to thee.
I pull the flax as golden as thy curl,
And sing me of the blossoms blue,
Their promise, like thine eyes to me.

Strumm, strumm!

'Tis such a merry tale I spinn.
Ah, wee one, croon unto, the honey bee
Who diggeth at the rose's heart.

Strumm, strumm!

My wheel shall sing to thee,
Heart-blossom mine.
The sunny mom
Doth bum with lovelilt, dear.
I fain would leave my spinning
To the spider climbing there,
And bruise thee, blossom, to my breast.

Strumm, strumm!

What fancies I do weave!
Thy dimpled band doth flutter, dear,
Like a petal cast adrift
Upon the breeze.

Strumm, strumm!

'Tis faulty spinning, dear.
A cradle built of thornwood,
A nest for thee, my bird.
I hear thy crooning, wee one,
And ah, this fluttering heart.

Strumm, strumm!

How ruthlessly I spinn!
My wheel doth wirr an empty song, my dear,
For tendrill nodding yonder
Doth nod in vain, my sweet;
And honey bee would tarry not
For thee; and thornwood cradle swayeth
Only to the loving of the wind!

Strumm, strumm!

My wheel still sings to thee,
Thou birdling of my fancy's realm!

Strumm, strumm!

An empty dream, my dear!
The sun doth shine, my bird;
Or should be fail, be shineth here
Within my heart for thee!

Strumm, strumm!

My wheel still Sings to thee.

Who would say that rhyme or measured lines would add anything to this unique song? It is filled with the images which are the essentials of true poetry, and it has the rhythm which sets the imagery to music and gives it vitality. "The tendrill tipped with sungilt," "the sunny morn doth hum with lovelilt," "thy dimpled hand doth flutter like a petal cast adrift upon the breeze"—these are figures that a Shelley would not wish to disown. There is a lightness and delicacy, too, that would seem to be contrary to our notions of the adaptiveness of blank verse. But these are technical features. It is the pathos of the song, the expression of the mother—yearning instinctive in every woman, which gives it value to the heart.

And yet there is a pleasure expressed in this song, the pleasure of imagination, which makes the mind's pictures living realities. In the poem which follows Patience expresses the feelings of the dreamer who is rudely awakened from this delightful pastime by the realist who sees but what his eyes behold:

Athin the even's hour,
When shadow purpleth the garden wall,
Then sit thee there adream,
And cunger thee from out the pack o' me.
Yea, speak thou, and tell to me
What 'tis thou hearest here.

A rustling? Yea, aright!
A murmuring? Yea, aright!
Ah, then, thou sayest, 'tis the leaves
That love one 'pon the other.
Yea, and the murmuring, thou sayest,
Is but the streamlet's hum.

Nay, nay! For wait thee.
Ayonder o'er the wall doth rise
The white faced Sister o' the Sky.
And lo, she beareth thee a fairies' wand,
And showeth thee the ghosts o' dreams.

Look thou! Ah, look! A one
Doth step adown the path!
The rustle?
'Tis the silken whisper o' her robe.
The hum?
The love-note o' her maiden dream.
See thee, ah, see!
She bendeth there,
And branch o' bloom doth nod and dance.
Hark, the note!
A robin's cheer?

Ah, Brother, nay.
'Tis the whistle o' her lover's pipe.
See, see, the path e'en now
Doth show him, tall and dark, aside the gate.

What! What! Thou sayest
'Tis but the rustle o' the leaves,
And brooklet's bumming o'er its stony path!
Then bush! Yea, bush thee!
Hush and leave me here!
The fairy wand hath broke, and leaves
Stand still, and note hath ceased,
And maiden vanished with thy word.

Thou, thou hast broke the spell,
And dream hath heard thy word and fled.
Yea, sunk, sunk upon the path,
They o' my dreams—slain, slain,
And dead with but thy word.
Ah, leave me here and go,
For Earth doth hold not
E'en my dreaming's wraith.

In previous chapters I have spoken of the wit and humor of Patience Worth. In only one instance has she put humor into verse, and that I have already quoted; but at times her poetry has an airy playfulness of form that gives the effect of humor, even though the theme and the intent may be serious. Here is an example:

Whiff, sayeth the wind,
And whiffing on its way, doth blow a merry tale.
Where, in the fields all furrowed and rough with corn,
Late harvested, close-nestled to a fibrous root,
And warmed by the sun that bid from night there—neath,
A wee, small, furry nest of root mice lay.
Whiff, sayeth the wind.

Whiff, sayeth the wind.
I found this morrow, on a slender stem,
A glory of the morn, who sheltered in her wine-red throat
A tiny spinning worm that wove the livelong day,—
Long after the glory had put her flag to mast—
And spun the thread I followed to the dell,
Where, in a gnarled old oak, I found a grub,
Who waited for the spinner's strand
To draw him to the light.
Whiff, sayeth the wind.

Whiff, sayeth the wind!
I blew a beggar's rags, and loving
Was the flapping of the cloth.
And singing on I went to blow a king's mantle 'bout his limbs,
And cut me on the crusted gilt.
And tainted did I stain the rose until she turned
A snuffy brown and rested her poor head
Upon the rail along the path.
Whiff, sayeth the wind.

Whiff, sayeth the wind.
I blow me 'long the coast,
And steal from out the waves their roar;
And yet from out the riffles do I steal
The rustle of the leaves, who borrow of the riffle's song

From me at summer-tide. And then
I pipe unto the sands, who dance and creep
Before me in the path. I blow the dead
And lifeless earth to dancing, tingling life,
And slap thee to awake at morn.
Whiff, sayeth the wind.

There is a vivacity in this odd conceit that in itself brings a smile, which is likely to broaden at the irony in the suggestion of the wind cutting itself on the crusted gilt of a king's mantle. Equally spirited in movement, but vastly different in character, is the one which follows:

Hi-ho, alack-a-day, whither going?
Art dawdling time away adown the primrose path
And wishing golden dust to fancied value?
Ah, catch the milch-dewed air, breathe deep
The clover-scented breath across the field,
And feed upon sweet-rooted grasses
Thou hast idly plucked.
Come, Brother, then let's on together.

Hi-ho, alack-a-day, whither going?
Is here thy path adown the hard-flagged pave,
Where, bowed, the workers blindly shuffle on;
And dumbly stand in gullies bound,
The worn, bedogged, silent-suffering beast,
Far driven past his due?
And thou, beloved, hast thy burden worn thee weary?
Come, Brother, then let's on together.

Hi-ho, alack-a-day, whither going?
Hast thou begun the tottering of age,
And doth the day seem over-long to thee?
Art fretting for release, and dost thou lack
The power to weave anew life's tangled skein?
Come, Brother, then let's on together.

The second line of this will at once recall Shakespeare's "primrose path of dalliance," and it is one of the rare instances in which Patience may be said to have borrowed a metaphor; but in the line which follows, "and wishing golden dust to fancied value," she puts the figure to better use than he in whom it originated. Beyond this line there is nothing especially remarkable in this poem, and it is given mainly to show the versatility of the composer, and as another example of her ability to present vivid and striking pictures.

Reference has been made to the love of nature and the knowledge of nature betrayed in these poems. Even in those of the most spiritual character nature is drawn upon for illustrations and symbols, and the lines are lavishly strewn with material metaphor and similes that open up the gates of understanding. This picture of winter, for example, brings out the landscape it describes with the vividness and reality of a stereoscope, and yet it is something more than a picture:

Snow tweaketh 'neath thy feet,
And like a wandering painter stalketh Frost,
Daubing leaf and lichen.
Where flowed a cataract
And mist-fogged stream, lies silvered sheen,
Stark, dead and motionless. I hearken
But to hear the she-e-e-e of warning wind,
Fearful lest I waken Nature's sleeping.
Await ye! Like a falcon loosed
Cometh the awakening.
Then returneth Spring
To nestle in the curving breast of yonder hill,
And sets to rest like the falcon seeketh
His lady's outstretched arm.

And here is another picture of winter, painted with a larger brush and heavier pigment, but expressing the same thought, that life doth ever follow death:

Dead, all dead!
The earth, the fields, lie stretched in sleep
Like weary toilers overdone.
The valleys gape like toothless age,
Besnaggled by dead trees.
The bills, like boney jaws whose flesh hath dropped,
Stand grinning at the deathy day.
The lily, too, hath cast her shroud

And clothed her as a brown-robed nun.
The moon doth, at the even's creep,
Reach forth her whitened bands and sooth
The wrinkled brow of earth to sleep.

Ah, whither flown the fleecy summer clouds,
To bank, and fall to earth in billowed light,
And paint the winter's brown to spangled white?
Where, too, have flown the happy songs,
Long died away with sighing
On the shore-wave's crest?
Will they take Echo as their Guide,
And bound from hill to hill at this,
The sleepy time of earth,
And waken forest song 'mid naked waste?
Ah, slumber, slumber, slumber on.
'Tis with a loving hand
He scattereth the snow,
To nestle young spring's offering,
That dying Earth shall live anew.

How different this from Thomson's pessimistic,

Dread winter spreads his latest glooms
And reigns tremendous o'er the conquered year.

This poem seemed to present unusual difficulties to Patience. The words came slowly and haltingly, and the indications of composition were more marked than in any other of her poems. The third line was first dictated "Like weary workmen overdone," and then changed to "weary toilers," and the eighteenth line was given: "On the shore-wavelet's breast," and afterwards altered to read "the shorewave's crest."

Possibly it was because the poet has not the same zest in painting pictures of winter that she has in depicting scenes of kindlier seasons, in which she is in accord with nearly all poets, and, for that matter, with nearly all people. Her pen, if one may use the word, is speediest and surest when she presents the beautiful, whether it be the material or the spiritual. She expresses this feeling herself with beauty of phrase and rhythm in this verse, which may be entitled "The Voice of Spring."

The streamlet under fernbanked brink
Doth laugh to feel the tickle of the waving mass;
And silver-rippled echo soundeth
Under over-hanging cliff.
The robin heareth it at morn
And steals its chatter for his song.
And oft at quiet-sleeping
Of the Spring's bright day,
I wander me to dream along the brooklet's bank,
And hark me to a song of her dead voice,
That lieth where the snowflakes vanish
On the molten silver of the brooklet's breast;
And watch the stream,
Who, over-fearful lest she lose the right
To ripple to the chord of Spring's full harmony,
Doth harden at her heart
And catch the song a prisoner to herself;
To loosen only at the wooing kiss
Of youthful Winter's sun,
And fill the barren waste with phantom spring.

Or, passing on to autumn, consider this apostrophe to a fallen leaf:

Ah, paled and faded leaf of spring agone,
Whither goest thou?
Art speeding
To another land upon the brooklet's breast?
Or art thou sailing to the sea, to lodge
Amid a reef, and, kissed by wind and wave,
Die of too much love?
Thou'lt find a resting place amid the moss,
And, ah, who knows!
The royal gem
May be thine own love's offering.
Or wilt thou flutter as a time-yellowed page,
And mould among thy sisters, ere the sun
May peep within the pack?

Or will the robin nest with thee
At Spring's awakening?

The romping brook
Will never chide thee, but ever coax thee on.
And shouldst thou be impaled
Upon a thorny branch, what then?
Try not a flight. Thy sisters call thee.
Could crocus spring from frost,
And wilt thou let the violet shrink and die?
Nay, speed not, for God hath not
A mast for thee provided.

Autumn, too, is the theme of this:

She-e-e! She-e-e! She-e-e-e!
The soughing wind doth breathe.
The white-crest cloud hath drabbed
At seasons late.
The trees drip leaf-waste
Unto the o'erloved blades aneath,
Who burned o' love, to die.

'Tis the parting o' the season.
Yea, and earth doth weep.
The mellow moon
Stands high o'er golded grain.
The cot-smoke
Curleth like to a loving arm
That reacheth up unto the sky.
The grain ears ope, to grin unto the day.
The stream hath laden with a pack o' leaves
To bear unto the dell, where bloom
Doth bide in waiting for her pack.
The stars do glitter cold, and dance to warm them
There upon the sky's blue carpet o'er the earth.

'Tis season's parting.
Yea, and earth doth weep.
The Winter cometh,
And be bears her jewels for the decking
Of his bride. A glittered crown
Shall fall 'pon earth, and sparkled drop

Shall stand like gem that flasheth
'Pon a nobled brow. Yea, the tears
Of earth shall freeze and drop
As pearls, the necklace o' the earth.
'Tis season's parting. Yea,
And earth doth weep.
'Tis Fall.

She does not confine herself to the Seasons in her tributes to the divisions of time. There are many poems which have the day for their subject, all expressive of delight in every aspect of the changing hours. There is a paean to the day in this:

The Morn awoke from off her couch of fleece,
And cast her youth-dampt breath to sweet the Earth.

The birds sent carol up to climb the vasts.
The sleep-stopped Earth awaked in murmuring.
The dark-winged Night flew past the Day
Who trod his gleaming upward way.
The fields folk musicked at the sun's warm ray.
Web-strewn, the sod, hung o'er o' rainbow gleam.
The brook, untiring, ever singeth on.

The Day hath broke, and busy Earth
Hath set upon the path o' hours.
Mute Night hath spread her darksome wing
And loosed the brood of dreams,
And Day hath set the downy mites to flight.
Fling forth thy dreaming hours! Awake from dark!
And hark! And hark! The Earth doth ring in song! '
Tis Day! 'Tis Day! 'Tis Day!

The close observer will notice in all of these poems that there is nothing hackneyed. The themes, the thoughts, the images, the phrasing, are almost if not altogether unique. The verse which follows is, I am inclined to believe, absolutely so:

Go to the builder of all dreams
And beg thy timbers to cast thee one.

Ah, Builder, let me wander in this land
Of softened shapes to choose.
My hand doth reach
To catch the mantle cast by lilies whom the sun
Hath loved too well. And at this morrow
Saw I not a purple wing of night
To fold itself and bask in morning light?
I watched her steal straight to the sun's
Bedazzled heart. I claim her purpled gold.
And watched I not, at twi-hours creep,
A heron's blue wing skim across the pond,
Where gulf clouds fleeted in a fleecy herd,
Reflected fair? I claim the blue and let
My heart to gambol with the sky-herd there.
At midday did I not then find
A rod of gold, and sun's flowers,
Bounded in by wheat's betasseled stalks?
I claim the gold as mine, to cast my dream.
And then at stormtide did I catch the sun,
Becrimsoned in his anger; and from his height
Did be not bathe the treetops in his gore?
The red is mine. I weave my dream and find
The rainbow, and the rainbow's end—a nothingness.

Almost equally weird is this "Birth of a Song ":

I builded me a harp,
And set asearch for strings.
Ah, and Folly set me 'pon a track
That set the music at a wail;
For I did string the harp
With silvered moon-threads;
Aye, and dead the notes did sound.
And I did string it then
With golden sun's-threads,
And Passion killed the song.
Then did I to string it o'er—
And 'twer a Jeweled string—
A chain o' stars, and lo,
They laughed, and sorry wert the song.

And I did strip the harp and cast
The stars to merry o'er the Night;
And string anew, and set athrob a string
Abuilded of a lover's note, and lo,
The song did sick and die,
And crumbled to a sweeted dust,
And blew unto the day.

Anew did I to string,
Astring with wail o' babe,
And Earth loved not the song.
I felled asorrowed at the task,
And still the Harp wert mute.
So did I to pluck out my heart,
And lo, it throbbed and sung,
And at the hurt o' loosing o' the heart
A song wert born.

That, however, is but a pretty play of fancy upon things within our ken, however shadowy and evanescent she may make them by her touch. But in the poem which follows she touches on the border of a land we know not:

I'd greet thee, loves of yester's day.
I'd call thee out from There.
I'd sup the joys of yonder realm.
I'd list unto the songs of them
Who days of me know not.
I'd call unto this hour
The lost of joys and woes.
I'd seek me out the sorries
That traced the seaming of thy cheek,
O thou of yester's day!

I'd read the hearts astopped,
That Earth might know the price
They paid as toll.
I'd love their loves,
I'd hate their hates,
I'd sup the cups of them;

Yea, I'd bathe me in the sweetness
Shed by youth of yester's day.
Yea, of these I'd weave the Earth a cloak—
But ah, He wove afirst!
They cling like petal mold, and sweet the Earth.
Yea, the Earth lies wrapped
Within the holy of its ghost.

'Tis but a drip o' loving", she said when she had finished this.

Nearly every English poet has a tribute to the Skylark, but
I doubt if there are many more exquisite than this:

I tuned my song to love and bate and pain
And scorn, and wrung from passion's beat the flame,
And found the song a wailing waste of voice.
My song but reached the earth and echoed o'er its plains.
I sought for one who sang a wordless lay,
And up from 'mong the rushes soared a lark.
Hark to his song!
From sunlight came his gladdening note.
And ah, his trill—the raindrops' patter!

And think ye that the thief would steal
The rustle of the leaves, or yet
The chilling chatter of the broolklet's song?
Not claiming as his own the carol of my heart,
Or listening to my plaint, he sings amid the clouds;
And through the downward cadence I but hear
The murmurings of the day.

One naturally thinks of Shelley's "Skylark" when reading this, and there are some passages in that celebrated poem that show a similarity of metaphor, such as this:

Sounds of vernal showers
On the twinkling grass;
Rain-awakened flowers;
All that ever was Joyous and clear and fresh
Thy music doth surpass.

And there is something of the same thought in the lines of Edmund Burke:

Teach me, O lark! with thee to greatly rise,
To exalt my soul and lift it to the skies;
To make each worldly joy as mean appear,
Unworthy care when heavenly joys are near.

But Patience nowhere belittles earthly joys that are not evil in themselves; nor does she teach that all earthly passions are inherently wrong: for earthly love is the theme of many of her verses.

Her expressions of scorn are sometimes powerful in their vehemence. This, on "War," for example:

Ah, thinkest thou to trick?
I fain would peep beneath the visor.
A god of war, indeed! Thou liest!
A masquerading fiend,
The harlot of the universe—
War, whose lips, becrimsoned in her lover's blood,
Smile only to his death-damped eyes!
I challenge thee to throw thy coat of mail!
Ah, God! Look thou beneath!
Behold, those arms outstretched!
That raiment over-spangled with a leaden rain!
O, Lover, trust her not!
She biddeth thee in siren song,
And clotheth in a silken rag her treachery,
To mock thee and to wreak
Her vengeance at thy hearth.
Cast up the visor's skirt!
Thou'lt see the snakey strands.
A god of war, indeed!
I brand ye as a lie!

Such outbreaks as this are rare in her poetry, but in her conversation she occasionally gives expression to anger or scorn or contempt, though, as stated, she seldom dignifies such emotions in verse. Love, as I have said, is her favorite theme in numbers, the love of God first and far foremost, and after that brother love and mother love. To the love of man for woman, or woman for man, there is seldom a reference in her poems, although it is the theme of some of her dramatic works. There is an exquisite expression of mother love in the spinning wheel lullaby already given, but for rapturous glorification of infancy, it would be difficult to surpass this, which does not reveal its purport until the last line:

Ah, greet the day, which, like a golden butterfly,
Hovereth 'twixt the night and morn;
And welcome her fullness—the hours
'Mid shadow and those the rose shall grace.
Hast thou among her hours thy heart's
Desire and dearest?
Name thou then of all
His beauteous gifts thy greatest treasure.
The morning, cool and damp, dark-shadowed
By the frowning sun—is this thy chosen?
The midday, flaming as a sword,
Deep-stained by noon's becrimsoned light
Is this thy chosen? Or misty startide,
Woven like a spinner's web and jeweled
By the climbing moon—is this thy chosen?
Doth forest shade, or shimmering stream,
Or wild bird song, or cooing of the nesting dove,
Bespeak thy chosen? He who sendeth light
Sendeth all to thee, pledges of a bonded love.
And ye who know Him not, look ye!
From all His gifts He pilfered that which made it His
To add His fullest offering of love.
From out the morning, at the earliest tide,
He plucked two lingering stars, who tarried
Lest the dark should sorrow.
And when the day was born,
The glow of sun-flush, veiled by gossamer cloud
And tinted soft by lingering night;
And rose petals, scattered by a loving breeze;
The lily's satin cheek, and dove cooes,
And wild bird song, and Death himself
Is called to offer of himself:
And soft as willow buds may be,

He claimeth but the down to fashion this, thy gift,
The essence of His love, thine own first-born.

In brief, the babe concentrates within itself all the beauties and all the wonders of nature. Its eyes, "two lingering stars who tarried lest the dark should sorrow," and in its face "the glow of sun flush veiled by gossamer cloud," "rose petals" and the "lily's satin cheek"; its voice the dove's coo. "From all His gifts He pilfered that which made it His"—the divine essence—"to add His fullest offering of love." This is the idealism of true poetry, and what mother looking at her own firstborn will say that it is overdrawn?

So much for mother-love. Of her lines on brotherhood I have already given example. In only a few verses, as I have said, does Patience speak of love between man and woman. The poem which follows is perhaps the most eloquent of these:

'Tis mine, this gift, ah, mine alone,
To paint the leaden sky to lilac-rose,
Or coax the sullen sun to flash,
Or carve from granite gray a flaming knight,
Or weave the twilight hours with garlands gay,
Or wake the morning with my soul's glad song,
Or at my bitterest drink a sweetness cast,
Or gather from my loneliness the flower
A dream amid a mist of tears.
Ah, treasure mine, this do I pledge to thee,
That none may peer within thy land; and only
When the moon shines white shall I disclose thee;
Lest, straying, thou should'st fade; and in the blackness
Of the midnight shall I fondle thee,
Afraid to show thee to the day.
When I shall give to Him, the giver,
All my treasure's stores, and darkness creeps upon me,
Then will I for this return a thank,
And show thee to the world.
Blind are they to thee, but ah, the darkness
Is illumined; and lo! thy name is burned
Like flaming torch to light me on my way.
Then from thy wrapping of love I pluck

My dearest gift, the memory of my dearest love.
Ah, memory, thou painter,
Who from cloud canst fashion her dear form,
Or from a stone canst turn her smile,
Or fill my loneliness with her dear voice,
Or weave a loving garland for her hair—
Thou art my gift of God, to be my comrade here.

Next to such love as this comes friendship, and she has put an estimate of the value of a friend in these words:

Of Earth there be this store of joys and woes.
Yea, and they do make the days o' me.
I sit me here adream that did I bold
From out the whole, but one, my dearest gift,
What then would it to be?
Doth days and nights
Of bright and dark make this my store?
Nay. Do happy hours and woes-tide, then,
Beset this day of me and make the thing I'd keep?
Nay. Doth metal store and jewelled string
Then be aworth to me? Nay. I set me here,
And dreaming, fall to reasoning for this,
That I would keep, if but one gift wert mine
Must bold the store o' all. Yea, must bold
The dark for light, yea, and bold the light for dark,
Aye, and bold the sweet for sours, aye, and bold
The love for Hate.
Yea, then, where may I to turn?

And lo, as I adreaming sat
A voice spaked out to me:
What ho! What ho! And lo, the voice of one, a friend!
This, then, shall be my treasure,
And the Earth part I shall bold
From out all gifts of Him.

Love of God, and God's love for us, and the certainty of life after death as a consequence of that love, are the themes of Patience's finest poetry, consideration of which is reserved for succeeding chapters.

Yet a taste of this devotional poetry will not be amiss at this point in the presentation of her works, as an indication of the character of that which is to come.

Lo, 'pon a day there bloomed a bud,
And swayed it at adance 'pon sweeted airs.
And gardens oped their greened breast
To shew to Earth o' such an one.
And soft the morn did woo its bloom;
And nights wept 'pon its check,
And mosses crept them 'bout the stem,
That sun not scoarch where it bad sprung.
And lo, the garden sprite, a maid,
Who came aseek at every day,
And kissed the bud, and cast o' drops
To cool the warm sun's rays.
And bud did hang it swaying there,
And love lept from the maiden's breast.

And days wore on; and nights did wrap
The bud to wait the morn;
And maid aseeked the spot.
When, lo, there came a Stranger
To the garden's wall,
Who knocked Him there
And bid the maiden come.

And up unto her heart she pressed her band,
And reached it forth to stay the bud's soft sway,
And lo, the sun hung dark,
And Stranger knocked Him there.
And 'twere the maid did step most regal to the place.
And barked, and lo, His voice aspoke.
And she looked upon His face,
And lo, 'twere sorry sore, and sad!
And soft there came His word
Of pleading unto her:
"O' thy garden's store do off offer unto me."
And lo, the maid did turn and seek her out the bud,
And pluck it that she bear it unto Him.

And at the garden's ope
He stood and waited her.
And forth her hand she held, therein the bud,
And lo, He took therefrom the bloom
And left the garden bare,
And maid did stand astripped
Of heart's sun 'mid her garden's bloom.
When lo, athin the wound there sunk
A warmpth that filled it up with love.
Yea, 'twere the smile o' Him, the price.

But she has given another form of poem which should be presented before this brief review of her more material verse is concluded, and it is a form one would hardly expect from such a source. I refer to the "poem of occasion." A few days before Christmas, Mrs. Curran remarked as she sat at the board: "I wonder if Patience wouldn't give us a Christmas poem." And without a moment's hesitation she did. Here it is:

I hied me to the glen and dell,
And o'er the heights, afar and near,
To find the Yule sprite's haunt.
I dreamt me it did bide
Where mistletoe doth bead;
And found an oak whose boughs
Hung clustered with its borrowed loveliness.
Ah, could such a one as she
Abide her in this chill?
For bleakness wraps the oak about
And crackles o'er her dancing branch.
Nay, her very warmth
Would surely thaw away the icy shroud,
And mistletoe would die
Adreaming it was spring.

I hied me to the holly tree
And made me sure to find her there.
But nay,
The thorny spines would prick her tenderness.
Ah, where then doth she bide?

I asked the frost who stood
Upon the fringed grasses 'neath the oak.
"I know her not, but I
Am ever bidden to her feast.
Ask thou the sparrow of the field.
He searcheth everywhere; perchance
He knoweth where she bides."

"Nay, I know her not,
But at her birthday's tide
I find full many a crumb
Cast wide upon the snow."

I found a chubby babe,
Who toddled o'er the ice, and whispered,
Did she know the Yule sprite's haunt?
And she but turneth solemn eyes to me
And wags her golden head.

I flitted me from house to shack,
And ever missed the rogue;
But surely she bad left her sign
To bid me on to search.
And I did weary of my task
And put my hopes to rest,
And slept me on the eve afore her birth,
Full sure to search anew at mom.

And then the morning broke;
And e'er mine eyes did ope,
I fancied me a scarlet sprite,
With wings of green and scepter of a mistletoe,
Did bid me wake, and whispered me
To look me to my heart.
Soft-nestled, warm,
I found her resting there.
Guard me lest I tell;
But, heart o'erfull of loving,
Thee'lt surely spill good cheer!

The following week, without request, she gave this New Year's poem, remarkable for the novelty of its treatment of a much worn theme:

The year hath sickened;
And dawning day doth show his withering;
And Death hath crept him closer on each hour. The crying hemlock shaketh in its grief.

The smiling spring hath hollowed it to age,
And golden grain-stalks fallen
O'er the naked breast of earth.
The year's own golden locks
Have fallen, too, or whitened,
Where they still do bold.

And do I sorrow me?
Nay, I do speed him on,
For precious pack be beareth
To the land of passing dreams.

I've bundled pain and wishing
'Round with deeds undone,
And packed the loving o' my heart
With softness of thine own;
And plied his pack anew
With loss and gain, to add
The cup of bitter tears I shed
O'er nothings as I passed.

Old year and older years—
My friends, my comrades on the road below—
I fain would greet ye now,
And bid ye Godspeed on your ways.

I watch ye pass, and read
The aged visages of each.

I love ye well, and count ye o'er
In fearing lest I lose e'en one of you.

And here the brother of you, every one,
Lies smitten!

But as dear I'll love him
When the winter's moon doth sink;
And like the watery eye of age
Doth close at ending of his day.
And I shall flit me through his dreams
And cheer him with my loving;
And last within the pack shall put
A Hope and speed him thence.

And bow me to the New.
A friend mayhap, but still untried.
And true, ye say?
But ne'er hath proven so!

Old year, I love thee well
And bid thee farewell with a sigh.

One who reads these poems with thoughtfulness must be impressed by a number of attributes which make them notable, and, in some respects, wholly unique. First of all is the absence of conventionality, coupled with skill in construction, in phrasing, in the compounding of words, in the application to old words of new or unusual but always logical meanings, in the maintenance of rhythm without monotony. Next is the absolute purity, with the sometimes archaic quality, of the English. It is the language of Shakespeare, of Marlowe, of Fletcher, of Jonson and Drayton, except that it presents Saxon words or Saxon prefixes which had already passed out of literary use in their time, while on the other hand it avoids nearly all the words derived directly from other languages that were habitually used by those great writers. There is rarely a word that is not of Anglo-Saxon or Norman birth. Nor are there any long words. All of these compositions are in words of one, two and three syllables, very seldom one of four—no "multitudinous seas incarnadine." Among the hundreds of words of Patience Worth's in this chapter there are only two of four syllables and less than fifty of three syllables. 95 percent of her works are in words of one and two syllables. In what other writing, ancient or modern, the Bible excepted, can this simplicity be found?

But the most impressive attribute of these poems is the weirdness of them, an intangible quality that defies definition or location, but which envelops and permeates all of them. One may look in vain through the works of the poets for anything with which to compare them. They are alike in the essential features of all poetry, and yet they are unalike. There is something in them that is not in other poetry. In the profusion of their metaphor there is an etherealness that more closely resembles Shelley, perhaps, than any other poet; but the beauty of Shelley's poems is almost wholly in their diction: there is in him no profundity of thought. In these poems there is both beauty and depth—and something else.

Chapter 5

THE PROSE

Word meeteth word, and at touch o' me, doth spell to thee.
—Patience Worth.

STRICTLY speaking, there is no prose in the compositions of Patience Worth. That which I have here classified as prose, lacks none of the essential elements of poetry, except a continuity of rhythm. The rhythm is there, the iambic measure which she favors being fairly constant, but it is broken by sentences and groups of sentences that are not metrical, and while it would not be difficult to arrange most of this matter in verse form, I am inclined to think that to the majority it will read smoother and with greater ease as prose. Nevertheless, as will be seen, it is poetry. The diction is wholly of that order, and it is filled with strikingly vivid and agreeable imagery. There is, however, this distinction: most of the matter here classed as prose is dramatic in form and treatment, and each composition tells a story—a story with a definite and well-constructed plot, dealing with real and strongly individualized people, and mingling humor and pathos with much effectiveness. They bring at once a smile to the face and a tear to the eye. They differ, too, from the poetry, in that they have little or no apparent spiritual significance. They are stories, beautiful stories, unlike anything to be found in the literature of any country or any time, but, except in the shadowy figure of The Stranger," they do not rise above the things of earth. That is not to say, however, that they are not spiritual in the intellectual or emotional sense of the word, as distinguished from the soul relation.

At the end of an evening a year and a half after Patience began her work, she said: "Thy hearth is bright. I fain would knit beside its glow and spinn a wordy tale betimes."

At the next sitting she began the "wordy tale." Up to that time she had offered nothing in prose form but short didactic pieces, such as will appear in subsequent chapters of this book, and the circle was lost in astonishment at the unfolding of this story, so different in form and spirit from anything she bad previously given.

Her stories are, as already stated, dramatic in form. Indeed they are condensed dramas. After a brief descriptive introduction or prologue, all the rest is dialogue, and the scenes are shifted without explanatory connection, as in a play. In the story of "The Fool and the Lady" which follows, the fool bids adieu to the porter of the inn, and in the next line begins a conversation with Lisa, whom be meets, as the context shows, at some point on the road to the tourney. It is the change from the first to the second act or scene, but no stage directions came from the board, no marks of division or change of scene, nor names of persons speaking, except as indicated in the context. In reproducing these stories, no attempt has been made to put them completely in the dramatic form for which they were evidently designed, the desire being to present them as nearly as possible as they were received; but to make them clearer to the reader the characters are identified, and shift of scene or time has been indicated.

The Fool and the Lady

AND there it lay, asleep. A mantle, gray as monk's cloth, 'its covering. Dim-glowing tapers shine like glow flies down the narrow winding streets. The sounds of early morning creep through the thickened veil of heavy mist, like echoes of the day afore. The wind is toying with the threading smoke, and still it clingeth to the chimney pot.

There stands, beyond the darkest shadow, the Inn of Falcon Feather, her sides becracked with sounding of the laughter of the king and gentlefolk, who barter song and story for the price of ale. Her windows sleep like heavy lidded eyes, and her breath doth reek with wine, last drunk by a merry party there.

The lamp, now blacked and dead, could boast to ye of part to many an undoing of the unwary. The roof, o'er-hanging and bepeaked, doth 'mind ye of a sleeper in his cap.

The mist now rises like a curtain, and over yonder steeple peeps the sun, his face washed fresh in the basin of the night. His beams now light the dark beneath the palsied stair, and rag and straw doth heave to belch forth its baggage for the night.

(Fool) "Eh, gad! 'Tis morn, Beppo. Come, up, ye vermin; laugh and prove thou art the fool's. An ape and jackass are wearers of the cap and bells. Thou wert fashioned with a tail to wear behind, and I to spin a tale to leave but not to wear. For the sayings of the fool are purchased by the wise. My crooked back and pegs are purses—the price to buy my gown; but better far, Beppo, to hunch and yet to peer into the clouds, than be as strong as knights are wont to be, and belly, like a snake, amongst the day's bright hours.

"Here, eat thy crust. 'Tis funny-bread, the earnings of a fool.

"I looked at Lisa as she rode her mount at yesternoon, and saw her skirt the road with anxious eyes. Dost know for whom she sought, Beppo? Not me, who, breathless, watched behind a flowering bush to hide my ugliness. Now laugh, Beppo, and prove thou art the fool's!

"But 'neath these stripes of color I did feel new strength, and saw me strided on a black beside her there. And, Beppo, knave, thou didst but rattle at thy chain, and lo, the shrinking of my dream!

"But we do limp quite merrily, and could we sing our song in truer measure—thou the mimic, and I the fool? Thine eyes hold more for me than all the world, since hers do see me not.

"We two together shall flatten 'neath the tree in yonder field and ride the clouds, Beppo, I promise ye, at after hour of noon.

"See! Tonio has slid the shutter's bolt! I'll spin a song and bart him for a sup."

(Tonio) "So, baggage, thou hast slept aneath the smell thou lovest best!"

(Fool) "Oh, morrow, Tonio. The smell is weak as yester's unsealed wine. My tank doth tickle with the dust of rust, and yet methinks thou would'st see my slattern stays to rattle like dry bones, to please thee. See, Beppo cryeth! Fetch me then a cup that I may catch the drops—or, here, I'll milk the dragon o'er thy door!"

(Tonio) "Thou scrapple! Come within. 'Tis he who loveth not the fool who doth hate his God."

(Fool) "I'm loth to leave my chosen company. Come, Beppo, his words are hard, but we do know his heart.

"A health to thee, Antonio. Put in thy wine one taste of thy heart's brew and I need not wish ye well.

"To her, Beppo. Come, dip and take a lick.

"Tonio, hast heard that at a time not set—is yet the tournament will be? Who think ye rides the King's lance and weareth Lisa's colors? Blue, Tonio, and gold, the heavens' garb—stop, Beppo, thou meddling pest! Antonio, I swear those bits of cloth are but patches I have pilfered from the ragheap adown the alleyway. I knew not they were blue. And this is but a tassel dropt from off a lance at yester's ride. I knew not of its tinselled glint, I swear!

"So, thou dost laugh? Ah, Beppo, see, be laughs! And we too, eh? But do we laugh the same? Come, jump! Thy pulpit is my bump. Aday, Antonio!"

(Antonio) "Aday, thou fool, and would I bad the wisdom of thy ape."

(On the Road to the Tournament.)

(Lisa) "Aday, fool!"

(Fool) "Ah, lady fair, hath lost the silver of thy laugh, and dost thee wish me then to fetch it thee?"

(Lisa) "Yea, jester. Thou speaketh wisely; for may I ripple laughter from a sorry heart? Now tease me, then."

(Fool) "A crooked laugh would be thy gift should I tease it with a crooked tale; and, lady, didst thee e'er behold a crooked laugh—one which holds within its crook a tear?"

(Lisa) "Oh, thou art in truth a fool. I'd bend the crook and strike the tear away."

(Fool) "Aye, lady, so thou wouldst. But thou hast ne'er yet found thy lot to bear a crook held staunch within His hand! Spring rain would be thy tears—a balm to buy fresh beauties. And the fool? Ah, his do dry in dust, e'en before they fall!"

(Lisa) "Pish, jester, thy tears would paint thy face to crooked lines, and thee wouldst laugh to see the muck. My heart doth truly sorry. Hast heard the King hath promised me as wages for the joust? And thee dost know who rideth 'gainst my chosen?"

(Fool) "Aye, lady, the crones do wag, and I do promise ye they wear their necks becricked to see his palfrey pass. They do tell me that his sumpter-cloth doth trail like a ladies' robe."

(Lisa) "Yea, fool, and pledge me thy heart to tell it not, I did broider at its hem a thrush with mine own tress—a song to cheer his way, a wing to speed him on."

(Fool) "Hear. Beppo, how she prates! Would I were a posey wreath and Beppo here a fashioner of song. We then would lend us to thy hand to offer as a token. But thou dost know a fool and ape are ever but a fool and ape. I'm off to chase thy truant laugh. Who cometh there? The dust doth rise like stormcloud along the road ahead, and 'tis shot with glinting. Oh, I see the mantling flush of morning put to shame by the flushing of thy cheek! See, he doth ride with helmet ope. Its golden bars do clatter at the jolt, and—but stop, Beppo, she beareth not! We, poor beggars, thee and me—an ape with a tail and a fool with a heart!

"See, Beppo, I did tear a rose to tatters but to fling its petals 'neath her feet. They tell me that his lance doth bear a ribband blue and a curling lock of gold—and yet he treads the earth! Let's then away!

The world may sorrow But the fool must laugh. 'Tis blessed grain
That hath no chaff.
To love an ape
Is but to ape at love.
I sought a band,
And found—a glove!

"Beppo, laugh, and prove thyself the fool's! I fain would feel the yoke, lest I step too high.

"Come, we'll seek the shelt'ring tree. I've in my kit a bit of curd. Thy conscience need not prick. I swear that Tonio, the rogue, did see me stow it there!

"Ah, me, 'tis such a home for fools, the 'earth. And they that are not fools are apes.

"I see the crowd bestringing 'long the road, and yonder clarion doth bid the riders come. Well, Beppo, do we ride? Come, chere, we may tramp our crooked path and ride astraddle of a cloud.

"She doth love him, then; and even now the born doth sound anew—and she the prize!

"I call the God above to see the joke that fate hath played; for I do swear, Beppo, that when he rides he carries on his lance-point this heart.

"I fret me here, but dare I see the play? Yea, 'tis a poor fool that loveth not his jest.

"I go, Beppo; I know not why, save I do love her so.

"I'll bear my hunch like a badge of His colors and I shall laugh, Beppo, shall laugh at losing. He loves me well, else why didst send me thee?

"The way seems over long.

"They parry at the ring! I see her veil to float like cloud upon the breeze.

"She sees me not. I wonder that she heareth not the thumping of my heart. My eyes do mist. Beppo, look thou! Ah, God, to see within her eyes the look of thine!

"They rank! And hell would cool my brow, I swear. Beppo, as thou lovest me, press sorely on my hump! Her face, Beppo, it swayeth everywhere, as a garden thick with bloom—a lily, white and glistening with a rain of tears. My heart hath torn asunder, that I know.

"The red knight now doth east! O Heaven turn his lance! "Tis put!

"And now the blue and gold! Wait, brother ape! Hold, in the name of God! Straight! 'Tis tie! Can I but stand? "I—ah, lady, he doth ride full well. May I but steady thee? My legs are wobbled but—my hand, dear lady, lest ye sink.

(" Beppo, 'tis true she seeth me!)

"Thy hand is cold. I wager you he wins. He puts a right too high. Thy thrush is singing; hear ye not his song? His wing doth flutter even now. Ah, he is fitting thee

"I do but laugh to feel the tickle of a feathering jest. An age before he puts! A miss! A tie! Ah, lady, should'st thee win I'll laugh anew and even then will laugh at what thee knowest not.

"The red knight! God weight his charger's hoof! (My God, Beppo, she did kiss my band!)

"He's off! Beppo, cling!"

(Lisa) "The fool! Look ye, the fool and ape! Oh heaven stop their flight! He's well upon them! Blind me, lest I die! He's charged anew, but missed! What, did his mantle fall? That shape that lieth! Come!"

(Lisa, to her knight) "So, thou, beloved, didst win me right! Where go they with the litter?"

(Knight) "The fool, my lady, and a chattering ape, did tempt to jest a charger in the field. We found them so. He lives but barely."

(Enter Fool upon litter.)

(Fool) "Aday, my lady fair! And hast thee lost the silver of thy laugh and bid me fetch it thee? The world doth hold but fools and lovers, folly sick."

(Lisa) "His eye grows misty. Fool, I know thee as a knave and love thee as a man."

(Fool) "'Tis but a patch, Beppo, a patch and tassel from a lance... but we did ride, eh? Laugh, Beppo, and prove thou art the fool's! I laugh anew, lest my friends should know me not. Beppo, I dream of new roads, but thou art there! And I do faint, but she... did kiss my hand... Aday L-a-d-y."

Very soon after the completion of this story Patience began another one, a Christmas story, a weird, mystical tale of medieval England, having for its central theme a "Stranger" who is visible only to Lady Marye of the Castle. The stranger is not described, nor does he speak a word, but he is presumedly the Christ. There are descriptions of the preparations for the Christmas feast at this lordly stronghold of baronial days, and the coarse wit of the castle servants and the drunken profanity of their master, "John the Peaceful," form a vivid contrast to the ethereal Lady Marye and the simple love of the herder's family at the foot of the hill. There are striking characters and many beautiful lines in this story, but it is not as closely woven nor as coherent in plot as the story of the fool and the lady.

The Stranger

'TWAS at white season o' the year, the shrouding o' spring and summerstide.

Steep, rugged, was the path, and running higher on ahead to turret-topped and gated castle o' the lordly state o' John the Peaceful, where Lady Marye whiled away the dragging day at fingering the regal.*

From sheltered niche she looked adown the hillside stretching 'neath. The valley was bestir. A shepherd chided with gentle word his flock, and gentle folk did speak o' coming Christ-time. Timon, the herder's hut, already hung with bitter sweets, and holly and fir boughs set to spice the air.

"Timon, man, look ye to the wee lambs well, for winter promiseth a searching night."

Thus spake Leta, who stands, her babe astride her hip.

"And come ye then within. I have a brew that of a truth shall tickle at thy funny bone. Bring then a bundle o' brush weed that we may ply the fire. I vow me thy boots are snow carts, verily!

"Hast seen the castle folk? And fetched ye them the kids? They breathe it here that the boar they roast would shame a heiffer. All of the sparing hours today our Leta did sniff her up the hill; nay, since the dawning she hath spread her smock and smirked.

"Leta, thou art such a joy! Thou canst wish the winter- painted bough to bloom, and like the plum flowers falls the snow. Fetch thee

* Regal. A small portable pipe organ used in the sixteenth and seventeenth centuries. It was played with one hand while the bellows was worked with the other.

a bowl and put the bench to table-side. Thy sire wouldst sup. Go now and watch aside the crib. Perchance thee'lt catch a glimpse o' heaven spilled from Tina's dream.

"Timon, man, tell me now the doings o' the day. I do ettle* for a spicey tale."

(Timon) "Well, be it so then, minx. I did fell the kids at sun-wake, and thee'lt find the skins aneath the cape I cast in yonder corner there. And I did catch a peep aslaunch** at mad Lady Marye, who did play the pipes most mournfully. They tell me she doth look a straining to this cot of ours. And what think ye, Leta? She doth only smile when she doth see our wee one's curls to glint. And ever she doth speak of him who none hath seen. 'Tis strange, think ye not?"

(Leta) "Nay, Timon, I full oft do pause and peer on high to see her at the summertide. Like a swan she bendeth, all white, amid her garden 'long the lake, and even 'tempts to come adown the path to us below. And ever at her heels the pea-fowl struts.

"She ne'er doth see my beckoning, but do I come with Tina at my breast she doth smile and wave and sway her arms a-cradle-wise.

"They tell, but breathlessly, that she doth sadly say the Stranger bideth here."

(Timon) "I'll pit my patch 'gainst purse o' gold, that 'Mad Marye' fitteth her as surely as 'Peaceful John' doth fit her sire. Thee knowest 'peace' to him is of his cutting, and 'piece' doth patch his ripping.

"They've bid a feast at Christ-night, and ye shouldst see the stir! I fain would see Sir John at good dark on that eve, besmeared with boar grease and soaked with ale, his mouth adrip with filth, and every peasant there who serves his bolts shall hit. And Lady Marye setteth like a lily under frost!

"Leta, little one, thine eyes do blink like stars beshadowed in a cloudy veil. Come, bend thy knee and slip away to dream!"

(Little Leta prays) "Vast blue above, wherein the angels hide; and moon, his lamp o' love; and cloud fleece white—art thou the wool to swaddle Him? And doth His mother bide upon a star-beam that leadeth her to thee? I bless Thy name and pray Thee keep my sire to watch full well his flock. And put a song in every coming day; my Tina's coo, and mother's song at eve. Goodnight, sweet night! I know He watcheth thee and me."

---

* Ettle. In this case, to have a strong desire.

** Aslaunch. Aslant or obliquely. As we would now say, "Out of the corner of the eye."

(Timon) "He heareth thee, my Leta. Watch ye the star on high. See ye, it winketh knowingly. God rest ye, blest."

(At the Castle.)

(Lady Marye) "And I the Lady Marye, o' the lord's

estate! Jana, fetch me a goblet that I drink."

(Jana) "Aye, lady. A wine, perchance?

(Lady Marye) "Nay, for yester thou didst fetch me wine, and I did cast it here upon the flags. Its stain thee still canst see. Shouldst thou fetch a goblet filled to brim with crystal drops, and I should cast it here, the greedy stone would sup it up, and where be then the stain? Think ye the stone then the wiser o' the two?

"I but loosed my fancy from its tether to gambol at its will, and they do credit me amiss. I weave not with strand upon a wheel. 'Tis not my station. Nay, I dally through the day with shuttle-cock and regal—a fitting play for yonder babe.

"Jana, peer ye to the valley there. Doth see the Stranger? He knocketh at the sill o' yonder cot.

"I saw him when the cotter locked the sheep to tap a straying ewe who lagged, and he did enter as the cotter stepped within—unbidden, Jana, that I swear—and now he knocketh there!"

(Jana) "Nay, lady, 'tis but a barish limb that reacheth o'er the door. The cotter heedeth not, ye see."

(Lady Marye) "I do see him now to enter, and never did be turn! Jana, look ye now! Doth still befriend a doubt?"

(Jana) "Come, lady, look! Sirrah John hath sent ye this, a posey, wrought o' gold and scented with sweet oils."

(Lady Marye) "Ah, Jana, 'tis a hateful sight to me—a posey I may keep! Why, the losing o' the blossom doth but make it dear!

"Stay! I know thee'lt say 'twas proffered with his love. But, Jana, thou hast much to learn. What, then, is love? Can I then sort my tinder for its building and ply the glass to start its flame? The day is o'er full now of ones who tried the trade. Nay, Jana, only when He toucheth thee and bids thee come and putteth to thy band His own doth love abide with thee.

"Come to the turret, then. I do find me whetted for a look within.

"How cool the eve! 'Tis creepy to the marrow. Look ye down the hillside there below. See ye the cotter's taper burning there? How white the night! 'Tis put upon the earth a mantling shroud, and sailing in the silver sky a fairy boat. Perchance it bringeth us the Babe.

"Jana, see'st thou the Stranger? He now doth count the sheep. Dare I trust him there? I see him fondling a lamb and he doth hold it close unto his breast."

(Jana) "Nay, lady, 'tis the shepherd's dog who skulketh now ahind the shelter wall."

(Lady Marye) "Ah, give me, spite o' this, the power to sing like Thine own bird who swayeth happily upon the forest bough and pours abroad his song where no man heareth him.

"Hear ye them below within the ball? They do lap at swine-broth. Their cups do clank. At morrow's eve they feast and now do need to stretch their paunches. Full often have I seen my ladye mother's white robe stained crimson for a jest, and oftener have I been gagged to swallow it. But, Jana, I do laugh, for the greatest jest is he who walloweth in slime and thinketh him a fish."

(Jana) "See, Lady Marye! This, thy mother's oaken chest, it still doth bear a scent o' her. And this, thy gown o' her own fashioning."

(Lady Marye) "Yea, Jana, and this o' her, a strand wound to a ball for mine own casting. And this! I tell thee, 'tis oft and oft she did press me to her own breast and chide me with her singing voice:' My Marye, 'tis a game o' buff, this living o' these days o' ours o' seeking happiness. When ye would catch the rogue he flitteth on.'

"See, these spots o' yellowed tears—the rusting of her heart away! Stay, Jana, I'll teach thee a trick o' tripping, for she full oft did say a heart could bide aneath a tripping.

"Thee shouldst curtsey so; and spread thy fan. 'Tis such a shield to bide ahind. Then shouldst thy heart to flutter, trip out its measure, so. See, I do laugh me now—nay, 'tis ne'er a tear, Jana, 'tis the mist o' loving! Doth see the moon hath joined the dance? Or, am I swooning? 'Tis fancy. See, the cotter's taper still doth flicker from the shutter. What's then amiss? The stranger, Jana! See! He entereth the shelter place! Come, I fear me lest I see too much? Lend me thy hand. I've played the jane-o-apes till the earth doth seem awry.

"Hear ye the wine-soaked song, and aye, the feed- drunkened? My sire, Jana, my sire! I do grow hateful of myself, but mark ye, at the setting o' the feast I do wage him war at words! A porridge pot doth brew for babes; I promise ye a full loaf. Do drop the curtain now, I weary me with reasoning."

(Morning at the Castle Gate.)

(Tito) "Aho, within! Thine eyes begummed and this the Christ-eve and mornin' come? Scatter! Petro, stand ahand! I do fetch ye sucklings agagged with apples red. Ye gad, my mouth doth slime! To whiff a hungerfull would make the sages wag."

(Petro) "Amorrow, Tito. Thee'lt wear thee white as our own Lady long afore ye e'en canst dip thy finger in the drip."

(Tito) "Pst! Petro, I did steal the brain and tung. Canst leave me have a peep now to the ball? Jesu! What a breeder o' sore bellies. I'd pay my price to heaven to rub Sir John a briskish rub with mullien o'er the back.

"They do tell me down below that trouble bideth Timon. His Tina layeth dull and Leta doth little but mumble prayer."

(Petro) "Tito, thee art a chanter of sad lays at this Christ- time. Go thou to the turret and play ye at the pipes. Put thee the sucklings to the kitchen, aside the fire dogs there. And Tito, thee'lt find a pudding pan ahind the brushbox. Go thee and lick it there!"

(To Sir John) "Aye, I do come, my lord. 'Tis but the sucklers come. I know not where in the castle she doth bide, but hark ye and ye'll surely hear the pipes."

(Sir John) "Bah! Damn the drivelling pipes! I do hear them late and early. 'Tis a fine bird for a lordly nest! Go, fetch her here! But no, I'd tweak her at a vaster sitting. Get thee, thou grunting swine! And take this as thy Christ-gift. I'd deal thee thrice the measure wert not to save these lordly legs. Here, fetch me a courser. I'd ride me to the hounds. And strip him of his foot cloth, that I do waste me not a blow. Dost like the smart? Or shall I ply it more? Thee'lt dance to tune, or damn ye, run from cuts!

"Ho, Timon, bow goes it with the brat? The world's o'erfull o' cattle now!"

(Timon) "Yea, sire, so did my Leta say when she did see thee come. 'Tis with our Tina as a bird behovered in the day. Aday, and God forgive thee."

(In Lady Marye's Chamber.)

(Lady Marye) "Jana, morn hath come. 'Tis Christ-tide and He not here! My limbs do fail, and how do I then to stand me thro' the day? The feast, the feast, yea, the feast! The day doth break thro' fog in truth!

"My mother's bridal robe! Go, Jana, fetch it me, and one small holly bough. Lend me a hand. I fain would see the cot.

"See thou! The sun doth love it, too, and chooseth him to rise him o'er its roof! Hath thee seen the herder yet to buckle loose the shelter place? And, Jana, did all seem well to thee? Nay, the Stranger, Jana! See, he still doth hold the lamb! 'My Marye, 'tis a game o' buff, this living o' these days o' ours.' In truth, 'tis put.

"Jana, I did dream me like a babe the night hours through; a dream so sweet, o' vast blue above wherein the angels hid, and I did see the Christ-child swaddled in a cloud; and Mary' maid of sorrows, led to him adown a silver beam.

"Then thee dost deem my fitful fancy did but play me false? Stay thou, my tears, and, heart o' me, who knoweth He doth watch o'er thee and me?

"Her robe! Ah, Fancy, 'tis thy right that thou art ever doubted. For thou art a conjurer, a trickster, verily. What chamming* joy didst thee then offer her?

"Thou cloud of billowed lace, a shield befitting her pure heart! And I the flowering of the bud! Hear me, all this o' her! I love thee well, and should the day but offer a bitter draft to quaff, 'tis but to whet me for a sweeter drink. And mother, heart o' me, hearken and do believe. I love my sire, Sir John.

"Come, Jana. Hear ye the carolers? Their song doth filter thro' my heart and lighten it. The snow doth tweak aneath their feet like pipes to 'company them. Cast ye a bit o' holly and a mistletoe.

"The feasters come to whet them with a pudding whiff. See, my sire doth ride him up the hill and o'er his saddle front a fallow deer.

Hear thee the cheering that he comes! Her loved, my Jana, and her heart doth beat through me!

"Christ-love to thee, my sire! Dost hear me here? And I do pledge it thee upon His precious drops caught by the holly tree. He seeth not, but she doth know!"

(Christmas Eve.)

(Jana) "My lady, who doth come a knocking at the door?

'Tis Petro, come to bid ye to the feast."

(Petro) "The candles are long since lit and Sirrah John hath wearyed him with jest. The feasting hath not yet begun, for he doth wait thee to drink a health to feasters in the hall."

(Lady Marye) "Yea, Petro, say unto my sire, the Lady Marye comes. And say ye more, she bids the feasters God- love. And say thee more,

* Obsolete form of "champing." Used here figuratively.

she doth bear the blessings of her Lady Mother who wisheth God's love to them all. And fetch ye candle trees to scores, and fetch the dulcimer and one who knocketh on its strings, and let him patter forth a lively tune, for Lady Marye comes.

"Jana, look ye once again to the valley there. The tapers burn not for Christ-night. Nay, a sickly gleam, and see, the Stranger, how he doth hold the lamb! And o'er his face a smile—or do my eyes beblur, and doth he weep?"

(Jana) "Nay, lady, all is dark. 'Tis but the whitish snow and shadow pitted by the tapers' light."

(Lady Marye) "Fetch me then my fan. I go to meet my Lord. Doth hear? Already they do play. I point me thus, and trip my heart's full measure."

(In the Hall.)

(Sir John) "So, lily-lip, thee'lt scratch! Thy silky paw hath claws, eh? Egad! A phantom! A ghoulish trick! My head doth split and where my tung? Get ye! Why sit like grinning asses! And where thy tungs? My God! What scent o' graves she beareth with that shroud!"

(Lady Marye) "God cheer, my lord, and doth my tripping suit thee well? These flags are but my heart and hers, and do I bruise them well for thee? Ah, aha! See, I do spread my fan. To shield my tears, ye think? Nay, were they to fall like Mayday's rain and thee wert buried 'neath a stone, as well then could'st thou see! And yet I love thee well. See thee, my sire, I pour this to thee!

"Look ye, good people at the feast; the boar is ready to slip its bones.

(Aside) "God, send Thy mantling love here to Thine own! For should I judge, when Thou I know dost love the saint and sinner as Thine own?

"To thee, my sire, to thee!

And gusted wind did flick the tapers out and they did hear her murmuring "The Stranger! He doth bid me come!"

And to this day they tell that Lady Marye cast the wine into a suckler's mouth and never did she drink!

"By all the saints! Do thee go and search!" Thus spake her sire, Sir John. And all the long night thro' the torches gleamed, but all in vain. And they do say that Sirrah John did shake him in a chilling and flee him to a friar, while still the search did last.

(In Timon's Cot.)

(Leta) "Timon, waken ye! Our Leta still doth court her dreams and I do weary me. The long night thro' the feasters cried them thro' the hills and none but Him could shield our Tina from their din.

"Take heart, my lad, I fear me yet to look within the crib. Hold thou my band, man. Nay, not yet! Come, waken Leta that she then do feed thy lambs."

(Timon) "Come, Leta, wake! The sun hath tipped the crown o' yonder hill and spread a blush adown her snow- white side."

(Leta) "Yea, sire. And Tina, how be she?" (Timon) "A fairy, sleeping, Tad."

(Leta) "Ah, sire, but I did dream the dark o' yesterday away. And, mother, she doth strain unto the sun! I see her eyes be-glistened. See, the frost-cart dumped beside our door, and look ye! he, the Frost man, put a cap upon the chimney pot. I'll fetch a brush and fan away his cloak. My Christ-gift, it would be my Tina's smile. She did know me not at late o' night; think ye it were the dark? Stay, sire! I'll cast the straw and put the sheep aright!" (Exit.)

(Timon) "My Leta, come! Thy Christgift bideth o'er our Tina's lips and she doth coo!"

(Leta) "Timon, call aloud, that she heareth thee. Leta! Leta! Little one! Dost hear thy sire to call? Why, what's amiss with thee? Thy staring eyes, my child! Speak thou!"

(Leta) "Sh-e-e-e! Sire. His mother's come! And, ah, my heart! All white she be an' crushed unto her breast a holly bough, and one white arm doth circle o'er a lamb! See, sire, the snow did drift it thro' and weave a fairy robe to cover her."

(Timon) "Who leaveth by the door; a stranger?"

(Leta) "Nay, He bideth here."

(Timon) "The Lady Marye, on my soul! Leta, drop ye here thy tears, for madness bideth loosed upon the earth! And shouldst

(Leta) "Nay, sire! Who cometh there?

And searchers there did find the Lady Marye, dead, amid the lambs and snow—a flowering o' the rose upon a bush o' thorn.

And hark ye! At the time when winter's blast doth sound, thee'lt hear the wailing o' the Lady Marye's pipes, and know the Stranger bideth o'er the earth.

The two dramatic stories presented here were but a paving of the way for larger work . *The Stranger* had been hardly completed when Patience announced, "Thee'lt sorry at the task I set thee next." And then she began the construction of a drama that in its delivery consumed the time of the sittings for several weeks, and it contained when finished some 20,000 words. It is divided into six acts, each with a descriptive prologue, and three of the acts have two scenes each, making nine scenes in all. It, like the two shorter sketches, is medieval in scene, and the pictures which it presents of the customs and costumes and manners of the thirteenth or fourteenth century (the period is not definitely indicated) are amazingly vivid. It has a somewhat intricate plot, which is carried forward rapidly and its strands skillfully interwoven until the nature of the fabric is revealed in the sixth act. This play is much more skillfully constructed in respect of stage technique than the two play lets that preceded it, and it could, no doubt, be produced upon the stage with perhaps a little alteration to adapt it to modern conditions. Some idea of its beauty, its sprightliness and its humor may be obtained from the prologue to the first act, which follows:

Wet earth, fresh trod.

Highway cut to wrinkles with cart wheels born in with o'erloading. A flank o' daisy flowers and stones rolled o'er in blanketing o' moss. Brown o' young oak-leaves shows soft amid the green. Adown a steep unto the vale, hedged in by flowering fruit and threaded through with streaming silver o' the brook, where rushes shiver like to swishing o' a lady's silk.

Moss-lipped log doth case the spring who mothereth the brook, and ivy hath climbed it o'er the trunk and leafless branch o' yonder birch, till she doth stand bedecked as for a folly dance.

Rat-a-tat! Rat-a-tat! Rat-a-tat! Sh-h-h-h!

From out the thick where hides the logged and mud- smeared shack.

Rat-a-tat! Rat-a-tat! Sh-h-h-h!

And hark ye, to the tanner's song!

Up, up, up! and down, down, down! A hammer to smite
And a hand to pound!
A maid to court,
And a swain to woo,
A heiffer felled
And I build a shoe!
A souse anew in yonder vat, And I'll buy my lady
A feathered hat!

The play then begins with the tanner and his apprentice, and the action soon leads to the royal castle, where the exquisite love story is developed, without a love scene. There is no tragedy in the story. It is all sentiment, and humor. And it is filled with poetry. Consider, for example, this description of Easter morn, from the prologue to the sixth act:

The earth did wake with boughs aburst. A deadened apple twig doth blush at casting Winter's furry coat, to find her naked blooms abath in sun. The feathered hosts, atuned, do carol, "He hath risen!" E'en the crow with envy trieth melody and soundeth as a brass; and listening, loveth much his song. Young grasses send sweet-scented damp through the hours of risen day. The bell, atoll, doth bid the village hence. E'en path atraced through velvet fields hath flowered with fringing bloom and jeweled drops, atempting tarriers. The sweet o' sleep doth grace each venturing face. The kine stand knee depth within the sillytittered brook, or deep in bog awallow. Soft breath ascent and lazy-eyed, they wait them for the stripping-maid.

The play is permeated with rich humor, and to illustrate this I give a bit of the dialogue between Dougal, the page, and Anne, the castle cook. To appreciate it one must know a little of the story. The band of the Princess Ermaline is sought by Prince Charlie, a doddering old rake, whom she detests, but whom for reasons of state she may be compelled to accept. However, she vows she will not speak while he is at court, nor does she utter a word, in the play, until the end of the last act. She has fallen in love with a troubadour, who has come from no one knows where, but who by his grace and his wit and his intelligence has made himself a favorite with all the castle folk. Anne has a roast on the spit, and is scouring a pot with sand and rushes, when Dougal enters the kitchen.

Dougal.—"Anne, goody girl, leave me but suck a bone. My sides have withered and fallen in, in truth."

Anne.—"Get ye, Dougal! Thy footprints do show them in grease like to the Queen's seal upon my floor!"

Dougal.—"The princess hath bidden me to stay within her call, but she doth drouse, adrunk on love-lilt o' the troubadour, and Prince of Fools (Prince Charlie) hath gone long since to beauty sleep. He tied unto his poster a posey wreath, and brushed in scented oils his beauteous locks, and sung a lay to Ermaline, and kissed a scullery wench afore he slept."

Anne.—"The dog! I'd love a punch to shatter him! And Ermaline hath vowed to lock her lips and pass as mute until his going."

Dougal.—"Yea, but eye may speak, for hers do flash like lightning, and though small, her foot doth fall most weighty to command.

"Yester, the Prince did seek her in the throne room. He'd tied his kerchief to a sack and filled it full o' blue-bells, and minced him 'long the halls astrewing blossoms and singing like to a frozen pump.

"Within the chamber, Ermaline did hide her face in dreading to behold him come, but at the door he spied the dear and bounded like a puppy 'cross the flags, apelting her with blooms and sputtering 'mid tee-hees. She, tho', did spy him first, and measured her his sight and sudden slipped her 'neath the table shroud. And be, Anne, I swear, sprawled him in his glee and rose to find her gone. And whacked my shin, the ass, acause I heaved at shoulders."

Anne.—"Ah, Dougal, "tis a weary time, in truth. Thee hadst best to put it back, to court thy mistress' whim. Good sleep, ye! And Dougal, I have a loving for the troubadour. Whence cometh be?"

Dougal.—"Put thy heart to rest, good Anne; he's but a piper who doth knock the taber's end and coaxeth trembling strings by which to sing. He came him out o' nothing, like to the night or day. We waked to hear him singing 'neath the wall."

Anne.—"Aye, but I do wag! For surely thee doth see how Ermaline doth court his song."

Dougal.—"Nay, Anne, 'tis but to fill an empty day."

When Patience had finished this she preened herself a little. "Did I not then spin a lengthy tale?" she asked. But immediately she began work upon another, a story of such length that it alone will make a book. It differs in many respects from her other works, particularly in the language, and from a literary standpoint is altogether the most amazing of her compositions. This, too, is dramatic in form, but scene often merges into scene without division, and it has more of the characteristics of the modern story. It is, however, medieval, but it is a tale of the fields, primarily, the heroine, Telka, being a farm lass, and the hero a field hand. Perhaps this is why the obscure dialectal forms of rural England of a time long gone by are woven into it. In this Patience makes an astonishingly free use of the prefix "a," in place of a number of prefixes, such as "be" and "with," now commonly used, and she attaches it to nouns and verbs and adjectives with such frequency as to make this usage a prominent feature of the diction. Let me introduce Telka in the words of Patience:

"Dewdamp soggeth grasses laid low aneath the blade at yester's harvest, and thistle-bloom weareth at its crown a jewelled spray.

"Brown thrush, nested 'neath the thick o' yonder shrub, hath preened her wings full long aneath the tender warmth o' morning sun.

"Afield the grasses glint, and breeze doth seeming set aflow the current o' a green-waved stream.

"Soft-footed strideth Telka, bare toes asink in soft earth and bits o' green acling, bedamped, unto her snowy limbs. Smocked brown and aproned blue, she seemeth but a bit o' earth and sky alight amid the field. Asplit at throat, the smock doth show a busom like to a sheen o' fleecy cloud aveiling o'er the sun's first flush.

"Betanned the cheek, and tresses bleached by sun at every twist of curl. Strong hands do clasp a branch long dead and dried, at end bepronged, and casteth fresh-cut blades to heap."

Such is Telka in appearance. "She seemeth but a bit o' earth and sky alight amid the field." Seemeth, yes, but there is none of the sky in Telka. She is of the earth, earthy, an intensely practical young woman, industrious, economical, but with no sense of beauty whatever, no imagination, no thought above the level of the ground. "I fashioned jugs o' clay," her father complained, "and filled with bloom, and she becracked their necks and kept the swill therein." Add to this a hot temper and a sharp tongue, and the character of Telka is revealed. Franco, the lover, on the other hand, is an artist and poet, although a field worker. He has been reared, as a foundling, by the friars in the neighboring monastery, and they have taught him something of the arts of mosaics and the illumination of missals. Between these two is a constant conflict of the material and the spiritual, and the theme of the story is the spiritual regeneration or development of Telka.

"See," says Franco, "Yonder way-rose hath a bloom! She be a thrifty wench and hath saved it from the spring."

Telka.—"I bate the thorned thing. Its barb hath pricked my flesh and full many a rent doth show it in my smock."

Franco.—"Ah, Telka, thine eyes do look like yonder blue and shimmer like to brooklet's breast."

Telka.—"The brooklet be bestoned, and muddied by the swine. Thy tung doth trip o'er pretty words."

Franco.—"But list, Telka, I would have thee drink from out my cup!"

Telka.—"Ah, show me then the cup."

And Telka's father, a wise old man, cautions Franco:

"Thee hadst best to take a warning, Franco. She be o' the field and rooted there; and thee o' the field, but reaped, and bound to free thee of the chaff by flailing of the world. She then would be to thee but straw and waste to cast awhither."

But an understanding of the nature of this strange tale and its peculiar dialect requires a longer extract. The "Story of the Judge Bush" will serve, better perhaps than anything else, to convey an idea of the characters of Telka and Franco, as well as to illustrate the language; and the episode is interesting in itself. The dialogue opens with Telka, Franco and Marion on their way to Telka's but. Marion is Telka's dearest friend, although one gets a contrary impression from Telka's caustic remarks in this excerpt; but unlike Telka, she can understand and appreciate the poetic temperament of Franco. To show her contempt for Franco's aspirations, Telka has taken his color pots and buried them in a dung- heap, and this characteristic act is the foundation of the "Story of the Judge Bush."

(Franco) "Come, we do put us to a-dry. 'Tis sky aweep, and 'tis a gray day from now.

I tell thee, Telka, we then put us to hearth, and spin ye shall. And thou, Marion, shalt bake an ash loaf and put o' apples for to burst afore the fire. 'Tis chill, the whine-wind o' the storm. We then shall spin a tale by turn; and Telka, lass, I plucked a sweet bloom for thee to wear. Thine eye hath softened, eh, my lass? Here, set thy nose herein and thou canst ne'er to think a tho't besoured."

(Telka) "Ah, 'tis a wise lad I wed, who spendeth o' his stacking hours to pluck weed, and thee wouldst have me sniff the dung-dust from their leaf. Do cast them whither, and 'pon thy smock do wipe thy hand. It be my fancy for to waste the gray hours aside the fire's glow,—but, Franco, see ye, the wee pigs asqueal! 'Tis nay liking the wet. Do fetch them hence. Here, Marion, east my cape about thee, since thou dost wear thy pettiskirt and Sabboth smock. Gad! Blue maketh thee to match a plucked goose. Thy skin already hath seamed, I vow. And, Marion, 'tis 'deed a flash to me thy tress be red! Should I to bear a red top I'd cast it whither."

(Franco) "Telka, Telka, drat thy barbed tung! Cast thou the bolt. Gad! What a scent o' browning joint!

(Telka) "Do leave me for to turn the spit that I may lick the finger-drip. Thy nose, Franco, doth trick thee. Thou canst sniff o' dung-dust and scoff at drip. Go, roll the apples o'er in yonder pile. They then would suit thee well!"

(Franco) "Telka, I bid thee to wash away such tunging. Here, I set them so. Now do I to fetch thy wheel. Nay, Marion, do cast thy blush. 'Tis but the Telka witch. Do thou to start thee at thy tale aspin."

(Telka) "Aye, Marion, thou then, since ne'er truth knoweth thee, thou shouldst ne'er to lack for story. Story do I say? Aye, or lie, 'tis brothers they be. And, Franco, do thou to spin, 'twill suit thy taste to feed 'pon maid's fare. I be the spinner o' the tale afirst. But, Franco, I fain would have thee fetch a pair o' barkers. Didst deem to fret me that thee dumped the twain aneath the stack? Go thou and fetch. 'Tis well that thee shouldst bed with swine lest thee be preening for a swan."

(Franco) "Ugh, Telka! Thou art like to a vat o' wine awork. Thou'lt fetch the swine do ye seek to company them."

(Telka) "So well, Polly, I do go, for 'tis swine o' worth amore than color daub. Set thee, since thou be wench."

(Franco) "Look ye, Telka, 'tis here I cast the cloak and show thee metal abared. Thou hast ridden 'pon a high nag for days, and I do kick his hock and set him at a limp. Do thou to clip thy words ashort or I do cast a stone athro' thy bubble."

(Telka) "Ah, Franco, 'tis nay meaning! Put here. Do spin thy tale, but do ye first to leave me fetch the wee-squeals. Then I do be a tamed dove. See ye?"

(Franco) "Away, then, and fetch thee back aburry." '(Exit Telka.)

(Franco) "Marion, 'tis what that I should put as path to tread? She be awronged but do I feed the fires, or put a stop?"

(Marion) "Franco, 'tis a pot and stew she loveth. Think ye to coax thy dream-forms from out the pot? Telka arounded and awrathed be like unto a thunder-storm, but Telka less the wrath and round, be Winter's dreary."

(Franco) "Not so, Marion, I shall then call forth the ghosts o' painted pots and touch the dreary abloom. Didst thou e'er to slit thy eye and view thro' afar? Dost thou then behold the motes? So, then, shall I to view the Telka maid. Whist! Here she be! Aback, Telka? Come, I itch for to spin a tale. Sit thee here and dry the wet sparkles from thy curls. List, do!

"'Twere a peddle-packer who did stroll adown the blade- strewn path along the village edge, abent. And brow-shagged eye did hide a twinkle-mirth aneath—"

"E-e-ek! E-e-e-k!"

(Telka) "Look, Franco, see they' e-e-e-k' do I to pull their tails uncurl!"

(Franco) "Do ye then wish thee, Telka, for to play upon their one-string lyre, or do I put ahead?"

"Bestrung, aborder o' the road, the cots send smoke- wreathes up to join the cloud. 'Twere sup-hour, and drip afrazzle soundeth thro' the doors beope, like to a water-cachit aslipping thro' dry leaf to pool aneath. Do I then put it clear?"

(Telka) "Yea, Franco, what hath he in his pack? I'd put a gander for a frock!"

(Marion) "On, Franco, thy tale hath a lilt."

(Franco) Awag-walk he weaveth to the door afirst-hand. The wee lads and lass do cluster 'bout the door, and twist atween their finger and thumb their smock-hem, or chew thereon. But he doth seem aloth to cast of pack or ope, and standeth at apeer to murmur—then to cast."

"E-e-e-k! E-e-e-k!

(Telka) "Nay, Franco, 'twere not my doing, I swear. 'Twere he who sat upon a firespark. Do baste! I hot for sight athin the pack."

(Franco) "What, Telka, thou awag and pig asqueak, and me the tail! Do put quiet!

The dame and sire do step them out from gray innards o' the hut, and pack-tipper beggeth for a mug o' porridge, and showeth o' the strand-bound pack. Wee lads and lass aquiver, tip-topple at a peep, and dame doth fetch the brew, but shaketh nay at offering o' gift, and spake it so: 'A porridge pot doth hold a mug, and one amore for he who bideth 'thout a brew. Nay, drink ye, and thank the morrow's sun. 'Tis stony path thee trod, and dust choketh. Do rest, and bide thee at our sill till weariness awarn away.'

"Think ye, Marion, that peddle-man did leave and cast not pence? What think ye, Telka?"

(Telka) "I did hear thee tell o' his fill, but tell thee o'fill o'pack."

(Franco) "A time, Telka. Nay, he did drink and left as price an ancient jug o' clay, and thick and o' a weight, to thank and wagweave hence."

(Telka) "Did he then to pack anew and off 'thout a peep?" (Franco) "Yea, and dark did yawn and swallow him. But morrow bringeth tale that peddle-packer had paid to each o' huts a beg, and what think ye? Left a jug where'er he supped!"

(Telka) "'Twere a clayster, and the morrow findeth him afollow for price, egh?"

(Franco) "Nay, Telka, not so. And jugs ashaken soundeth like to a wine; but atip did show nay drop. Marion, do tweak the Telka—she be aslumber."

(Marion) "Wake thee, Telka, the jugs be now to crack." (Telka) "Nay, 'tis a puddle o' a tale—a packster and a strand-bound pack, aweary."

(Franco) "But list thee! For 'twere eve that found the dames awag. For tho' they set the jugs aright, there be but dust where they did stand. Yea, all, Telka maid, save that the peddle-man did give to dame at first hand. The gabble put it so, that 'twere the porridge begged that dames did fetch but for a hope o' price, where jugs ashrunk."

(Telka) "But 'twere such a scurvey, Franco! I wage the jug aleft doth leak. What think ye I be caring 'bout jug or peddlepacker?"

(Marion) "Snip short thy word, Telka. Leave Franco for to tell. I be aprick for scratch to case the itch o' wonder. On, lad, and tie the ends o' weave-strand."

(Franco) "'Tis told the dame did treasure o' the jug, and sire did shew abroad the wonder, and all did list unto the swish o' 'nothing wine,' and thirsted for asup, and each did tip its crook'd neck and shake, but ne'er a drop did slip it through. And wonder, Marion, the sides did sweat like to a damp within! So 'twere. The townsmen shook awag their heads and feared the witch-work or the wise man's cunger, and they did bid the sire to dig a pit and put therein the jug."

(Telka) "'Twere waste they wrought, I vow, for should ye crack away its neck 'twould then be fit for holding o' the swill. There be a pair ahind the stack."

(Franco) "Nay, Telka, not as this, for they did dig a pit and plant jug therein, and morrow showed from out the fresh-turned earth a bush had sprung, and on its every branch a bud o' many colored hue alike to rainbow's robe. And lo, the dames and sires did cluster 'bout, and each did pluck a twig aladen with the bud, but as 'twere snapped, what think ye? There be in the hand a naught—save when the dame who asked not price did pluck. And 'tis told that to this day the townsmen fetch unto the bush and force apluck do they make question o' their brotherman. And so 'tis with he who fashions o' the rainbow's robe a world to call his own, and fetcheth to the grown bush his brother for to shew, and be seeth not, 'tis so be judge."

(Telka) "O, thou art a story-spinner o' a truth, and peddle- packer too, egh? And thou dost deem that thou hast planted o' thy pot to force thy bush by which ye judge. Paugh! Thou art a fool, Franco, and thy pots o' color be not aworth thy pains. So thou dost think then I be plucking o' naught aside thy bush. Well, I do tell thee this. Thy pots ne'er as the jug shall spring. Nay, for morn found me adig, and I did cast them here to the fire, afearing they should haunt."

(Franco) "'Tis nuff, Telka, I leave them to the flame. But thou shouldst know the bush abud doth show in every smouldering blaze."

(Telka) "See, Franco, I be yet neck ahead, for I do spat upon the flame and lo, thy bush be naught!"

(Franco) "Aye, 'tis so, but there be ahid a place thou ne'er hast seen. Therein I put what be mine own—the love for them. Thou art a butterfly, Telka, abeating o' thy wing upon a thistle-leaf. Do hover 'bout the blooms thou knowest best and leave dreambush and thistle-leaf."

It is a remarkable story. Many lines are gems of wit or wisdom or beauty, and it contains some exquisite poetry. There are many characters in it, all of them lovable but Telka, and she becomes so ere the end.

A curious and interesting fact in this connection is that after beginning this story Patience used its peculiar form of speech in her conversation and in her poems. Previously, as I have pointed out, there was a natural and consistent difference between her speech and her writings, and it would seem that in this change she would show that she is not subject to any rules, nor limited to the dialect of any period or any locality. Scattered through this present volume are poems, prose pieces and bits of her conversation, in which the curious and frequent use of the prefix a-, the abbreviation of the word "of" and the strange twists of phrase of the Telka story are noticeable. All of these were received after this story was begun.

But there is another form of prose composition that Patience has given to us. While she is writing a story she does not confine herself to that work, but precedes or follows it with a bit of gossip, a personal message, a poem or something else. Sometimes she stops in the midst of her story to deliver something entirely foreign to it that comes into her mind. During one week, while "Telka" was being received, she presented three parables, all in the peculiar language of that story. I reproduce them here and leave it to the reader to ponder over their meaning.

"Long, yea, long agone, aside a wall atilt who joined unto a brother-wall and made atween a gap apoint abacked, there did upon the every day, across-leged, sit a bartmaker, amid his sacks aheaped. And ne'er a buy did tribesmen make. Nay, but 'twere the babes who sought the bartman, and lo, he shutteth both his eyes and babes do pilfer from the sacks and feed thereon, till sacks asink. And still at crosslegs doth he sit.

"Yea, and days do follow days till Winter setteleth 'pon his locks its snow. Aye, and lo, at rise o' sun 'pon such an day as had followed day since first he sat, they did see that he had ashrunked and they did wag that 'twere the wasting o' his days at sitting at crossleg.

"And yet the babes did fetch for feast and wert fed. Till last a day did dawn and gap ashowed it empty and no man woed; but babes did sorry 'bout the spot 'till tribesmen marveled and fetched alongside and coaxed with sweets their word. But no man found answer in their prate. And they did ope remaining sacks and lo, there be anaught save dry fruit, and babes did reach forth for it and wert fed, and more, it did nurture them, and they went forth alater to the fields o' earth astrengthened and fed 'pon—what, Brother? List ye. 'Pon truth."

"There be aside the market's place a merchant and a brother merchant. Aye, and one did put price ahigh, and gold aclinketh and copper groweth mold atween where he did store. And his brother giveth measure full and more, for the pence o' him who offereth but pence, at measure that runneth o'er to full o' gold's price.

"And lo, they do each to buy o'herds, and he who hath full price buyeth but the shrunk o' herd, and he who hath little, buyeth the full o' herd. And time maketh full the sacks o' him who hoardeth gold, and layeth at aflat the sacks o' him who maketh poor price. And lo, he who hath plenty hoardeth more, and he who had little buyed o' seed and sowed and reaped therefrom. And famine crept it nearer and fringed 'pon the land and smote the land o' him who asacketh o' gold and crept it 'pon the land o' him o' pence.

"And herds did low o' hunger and he who hath but gold hath naught to feed thereon. For sacks achoked 'pon gold. And he who had but pence did sack but grain and grass and fed the herd. And lo, they fattened and did fill the emptied sacks with gold, while he who hath naught but gold did sick, and famine wasted o' his herd and famine's sun did rise to shine 'pon him astricken 'pon gold asacked."

"There wert a man and his brother and they wrought them unalike. Yea, and one did fashion from wood, and ply till wonderwork astood, a temple o' wood. And his brother fashioneth o' reeds and worketh wonder baskets. And he who wrought o' wood scoffeth.

And the tribesmen make buy o' baskets and wag that 'tis a- sorry wrought the temple, and spake them that the Lord would smite, and lay it low. For he who wrought did think him o' naught save the high and wide o' it, and looked not at its strength or yet its stand 'Pon earth. And they did turn the baskets 'bout and put to strain, and lo, they did hold. And it were the tribesmen, who shook their beads and murmured, 'Yea, yea, they be a goodly.'

"So 'tis; he who doth fashion from wood o' size doth prosper not, and he who doth fashion o' reed and small, doth thrive verily."

These are all somewhat cryptic, although their interpretation is not difficult, but that which follows on the magic of a laugh needs no explanation. "I do fashion out a tale for babes," said Patience, when she presented this parable of the fairy's wand, and in it she gives expression to another one of her characteristics, one that is intensely human, the love of laughter, which she seems to like to hear and often to provoke.

Lo, at a time thou knowest not, aye, I, thy handmaid, knowest not, there wert born unto the earth a babe. And lo, the dame o' this babe wert but a field's woman. And lo, days and days did pass until the fullness of the babe's days, and it stood in beauty past word o' me.

"Yea, and there wert a noble, and he did pass, and lo, his brow was darked, and smile bad forsook his lips. And he came unto the cot and there stood the babe, who wert now a maid o' lovely. And he spaked unto her and said:

"'Come thou, and unto the lands of me shall we make way. Thou art not o' the fields, but for the nobles.'

"And she spake not unto his word. And lo, the mother of the babe came forth and this man told unto her of this thing, that her babe wert not of the field but for the nobled. And, at the bidding of the noble, she spake, yea, the maid should go unto his lands.

"And time and time after the going, lo, no word came unto the mother. And within the lands of the noble, the maid lived, and lo, the days wert sorry, and the paths held but shadows, and nay smiles shed gold unto the hours. And she smiled that this noble did offer unto her much of royal stores. Yea, gems, and gold, and all a maid might wish, and she looked in pity unto the noble and spake:

"'What hast thou? Lo, thou hast brought forth of thy store and given unto me, and what doth it buy? Thy lips are ever sorry and thy hours dark. Then take thou these gifts and keep within such an day as thine, for, hark ye, my dame, the field's woman, hath given unto me that which setteth at a naught thy gifts; for hark ye: mid thy dark o' sorry I shall spill a laugh, and it be a fairies' wand, and turneth dust to gold.'

"And she fled unto the sun's paths of the fields.

"Verily do I to say unto thee, this, the power of the fairies' wand, is thine, thy gift of thy field—mother, Earth. Then cast out that which earth-lands do offer unto thee and flee with thy gift."

It is somewhat difficult to select an ending for this chapter on the prose of Patience: the material for it is so abundant and so varied, but this "Parable of the Cloak" may perhaps form a fitting finish:

"There wert a man, and lo, he did to seek and quest o' sage, that which he did mouth o'ermuch. And lo, he did to weave o' such an robe, and did to clothe himself therein. And lo, 'twer sun ashut away, and cool and heat and bright and shade.

"And lo, still did he to draw 'bout him the cloak, and 'twer o' the mouthings o' the sage. And lo, at a day 'twer sent abroad that Truth should stalk 'pon Earth, and man, were he to look him close, shouldst see.

"And lo, the man did draw 'bout him the cloak, and did to wag him 'Nay' and 'Nay, 'twer truth the sages did to mouth and I did weave athin the cloak o' me.'

"And then 'twer that Truth did seek o' Earth, and she wert clad o' naught, and seeked the man, and begged that he would cast the cloak and clothe o' her therein. And lo, he did to draw him close the cloak, and hid his face therein, and wag him 'Nay,' he did to know her not.

"And lo, she did to fetch her unto him athrice, and then did he to wag him still a 'Nay! Nay! Nay!' And lo, she toucheth o' the cloth o' sage's mouths and it doth fall atattered and leave him clothed o' naught, and at a wishing. And he did seek o' Truth, aye, ever, and when he did to find, lo, she wagged him nay, and nay, and nay."

Chapter 6

CONVERSATIONS

This be bread. If man knoweth not the grain from which 'twer fashioned, what then? 'Tis bread. Let man deny me this.
—Patience Worth.

BUT after all, perhaps the truest conception of the character and versatility of Patience can be acquired from her "conversations." The word "conversation" I here loosely apply to all that comes from her in the course of an evening, excepting the work on her stories. The poems and parables are usually woven into her remarks with a sequence that suggests extemporaneous production for the particular occasion, although as a rule they are of general application. Almost invariably they are brought out by something she or someone else has said, or as a tribute, a lesson or a comfort to some person who is present. Her songs, as she calls her poems, are freely given, apparently without a thought or a care as to what may become of them. They seem to come spontaneously, without effort, with no pause for thought, no groping for the right word, and to fall into their places as part of the spoken rather than the written speech. So it is that the term "conversation" in this connection is made to include much that ordinarily would not fall within that designation.

One of the pleasures of an evening with Patience is the uncertainty of the form of the entertainment. Never are two evenings alike in the general nature of the communications. She adapts herself to circumstances and to the company present, serious if they are bent on serious subjects, merry if they are so; but seldom will the serious escape without a little of the merry, or the merry without a little of the serious. Sometimes her own feelings seem to have an influence. Always, however, she is permitted to take her own course, except in the case of a formal examination, to which she readily responds if conducted with respect. She may devote the evening largely to poetry, possibly varying the themes, as on one evening when she gave a nature poem, one of a religious character, a lullaby, a humorous verse and a prayer, interspersed with discussion. She may talk didactically with little or no interruption. She may submit to a catechism upon religion, philosophy, philology, or any subject that may arise. She may devote an evening to a series of little personal talks to a succession of sitters, or she may elect just to gossip. "I be dame," she says, and therefore not averse to gossip. But rarely will she neglect to write something on whatever story she may have in band. She speaks of such writing as "weaving." "Put ye to weave," she will say, and that means that conversation is to stop for a time until a little real work is accomplished.

The conversations which follow are selected to illustrate the variety of form referred to, as well as to introduce a number of interesting statements that throw light on the character of the phenomena,

Upon a certain evening the Currans had two visitors, Dr. and Mrs. W. With Dr. W. and Mrs. C. at the board and Mrs. W. leaning over it, Patience began:

Ah, hark! Here abe athree; yea, love, faith and more o' love! Thee hast for to hark unto word I do put o' them, not ye."

And then she told this tale of the Mite and the Seeds:

"Hark! Aneath the earth fell a seed, and lay aside a Mite, a winged mite, who hid from cold. Yea, and the Mite knew o' the day o'er the Earth's crust, and spake unto the Seed, and said:

"'The hours o' day show sun and cloud, aye, and the Earth's crust holdeth grass and tree. Aye, and men walk 'pon the Earth.'

Aye, and the Seed did say unto the Mite:

"' Nay, there be a naught save Earth and dark, for mine eye hath not beheld what thou tellest of.'

Yea, and the Mite spake it so:

' 'Tis dark and cold o'er the crust o' Earth, and thou and me awarm and close ahere.'

But the Seed spake out: 'Nay, this be the time I seek me o'er the Earth's crust and see the Day thou tellest of.'

"And lo, be sent out leaf, and reached high. And lo, when the leaf had pushed up from 'neath the crust, there were snow's cut and cold, and it died, and knew not the Day o' the Mite: for the time was not riped that be should seek unto new days.

"And lo, the Stalk that had sent forth the Seed, sent forth amore, and lo, again a one did sink aside the Mite. And he spake to it of the Day o' Earth and said: 'Thy brother sought the Day, and it wert not time, and lo, he is no more.'

"And be told of the days of Earth unto the seed, and it spaked unto him and said: 'This day o' thee meaneth naught to me. Lo, I shall spring not a root, nor shall I to seek me the days o' Earth. Nay, I shall lay me close and warm.'

"And e'en though the Mite spake unto the Seed at the time when it wert ripe that it should seek, lo, it lay, and Summer's tide found it a naught, for it feeded 'pon itself, and lo, wert not.

And at a later tide did a seed to fall, and it harked unto the Mite and waited the time, and when it wert riped, lo, it upped and sought the day. And it wert so as the Mite had spaked. And the Seed grew into a bush.

"And lo, the winged Mite flew out: for it had brought a brother out o' the dark and unto the Day, and the task wert o'er.

"These abe like unto them who seek o' the Words o' me.

"Now aweave thou."

Patience then wrote about two hundred words of a story, after which Mrs. W. inquired of Mr. C.:

"Don't you ever try to write on the board?

To which he replied facetiously, "No, I'm too dignified."

Patience.—"Yea, he smirketh unto swine and kicketh the nobles."

Then seeming to feel that the visitors were wanting something more personal than the "Tale" she said:

"Alawk, they be ahungered, and did weave a bit. Then hark. Here be.

"What think ye, man? They do pucker much o'er the word o' me, and spat forth that thou dost eat and smack o' liking. Yea, but hark! Who shed drop for Him but one o' His, yea, the Son o' Him? Think ye this abe the pack o' me? Nay, and thou and thou and thou shalt shed drops in loving for the pack, for it be o' Him. Now shall I to sing:

How doth the Mise-man greed,
And lay unto his store,
And seek him out the pence of Earth,
Wherein the hearts do rust?

How doth the Muse-man greed,
And seek him o' the Day,
And word that setteth up a wag—
While hearts o' Earth are filthed?

How doth the See-man greed?
Yea, and how be opeth up his eye,
And seeth naught and telleth much—
While hearts of earth are hurt.

How doth the Good-man greed,
Who dealeth o' the Word?
He eateth o' its flesh and casts but bone,
While hearts o' Earth are woed.

How doth the Man-man greed?
He eateth o' the store, yet holdeth ope
His bands and scattereth o' bread
And hearts o' Earth are fed.
This then abe, and yet will be
Since time and time, and beeth ever."

As soon as this was read, she followed with another song:

Drink ye unto me.
Drink ye deep, to me.
Yea, and seek ye o' the Brew ye quaff,
For this do I to beg.
Seek not the wine o' Summer's sun,
That bid 'mid purpled vine,
And showeth there amid the Brew

Thou suppest as the Wine.
Seek not the drops o' pool,
Awarmed aneath the sun,
And idly lapping at the brink
Of mosses' lips, to sup.

Seek not o' vintage Earth doth bold.
Nay, unto thee this plea shall wake
The Wine that thou shouldst quaff.
For at the loving o' this heart
The Wine o' Love shall flow.
Then drink ye deep, ah, drink ye deep,
And drink ye deep o' Love.

Yea, thine unto me, and mine to thee."

After which she explained:

"I did to fashion out a brew for her ayonder and him ahere. And they did eat o' it. Yea, for they know o' Him and know o' the workings o' Him and drinked o' the love o' me as the love o' Him. Yea, and hark, there abe much athin this pack for thee."

This, it will be observed, is rather a discourse than a conversation, and it is often so, Patience filling the evening with her own words; not as exclusively so, however, as this would indicate: for there is always more or less conversation among the party, which it would profit nothing to reproduce.

The next sitting is somewhat more varied. There were present Dr. X., a teacher of anatomy, Mrs. X., Mrs. W. and Miss B. Dr. X. sat at the board with Mrs. Curran:

Patience.—"Eh, gad! Here be a one who taketh Truth unto him and setteth the good dame apace that she knoweth not the name o' her. I tell thee 'tis he who knoweth her as a sister, and telleth much o' her, and naught he speaketh oft holdeth her, and much he speaketh holdeth little o' her, and yet ever he holdeth her unto him. He taketh me as truth, yea, he knoweth be taketh naught and buildeth much, and much and buildeth little o' it. I track me unto the door o' him and knock and he heareth me."

This, of course, referred to Dr. X. and his work, and it aroused some discussion, after which Patience asked, "Would ye I sing?" The answer being in the affirmative, she gave this little verse, also directed to Dr. X.:

Out 'pon the sea o' learning,
Floateth the barque o' one aseek.
Out 'pon troubled waters floateth the craft,
Abuilded staunch o' beams o' truth.
And though the waves do beat them high
And wash o'er and o'er the prow,
Fear thee not, for Truth saileth on.
Set thy beacon, then, to crafts not thine,
For thou hast a light for man.

There, thou knowest me. I tell thee I speak unto him who hath truth for his very own. Set thee aweave."

The sitters complied and received about six hundred words of the story, after which Mrs. X. took the board, remarking as she did so that she was afraid, which elicited this observation from Patience:

"She setteth aside the stream and seeth the craft afloat and be at wishing for to sail, and yet she would to see her who steereth."

Mrs. X. gave up her place to Miss B., a teacher of botany, to whom Patience presented this tribute:

"The eye o' her seeth but beauties and shutteth up that which showeth darked, that that not o' beautie setteth not within the see o' her. Yea, more; she knoweth bow 'tis the dark and what showeth not o' beauty, at His touching showeth lovely for the see o' her.

"Such an heart! Ah, thou shouldst feast hereon. I tell thee she giveth unto multitudes the heart o' her; and such as she dealeth unto earth, earth has need for much. She feasteth her 'pon dusts and knoweth dust shall spring forth bloom. Hurt hath set the heart o' her, and she hath packed up the hurt with petals."

Patience then turned her attentions again to Dr. X. "He yonder," she said, "hath much aneath his skull's-cap that he wordeth not."

Thus urged, Dr. X. inquired:

"Does Patience prepare the manuscript she gives in advance? It rather seems that she reads the material to Mrs. Curran."

"See ye," cried Patience, "he hath spoke a thing that set aneath his skull's-cap!" And then, in answer to his question:

"She who afashioneth loaf doth shake well the grain-dust that husks show not. Then doth she for to brew and stir and mix, else the loaf be not afit for eat."

By grain-dust she means flour or meal, and she uses the word brew in its obsolete sense of preparation for cooking. The answer may be interpreted that she arranges the story in her mind before its dictation, and as to her formal work she has said many things to indicate that such is her method. Dr. X. then asked:

"Are these stories real happenings?

To which Patience replied:

"Within the land o' here [her land] be packed the days o' Earth, and thy day hath its sister day ahere, and thy neighbor's day and thy neighbor's neighbor's day. And I tell thee, didst thou afashion tale thou couldst ne'er afashion lie, for all thou hast athin thy day that thy put might show from the see o' thee hath been; at not thy time, yea, but it hath been."

"Then," asked Dr. X., "should you have transmitted through one who spoke another language you would have used their tongue?

Patience answered:

"I pettiskirt me so that ye know the me of me. Yea, and I do to take me o' the store o' her that I make me word for thee."

"Pettiskirt" is a common expression of hers to mean dress, in either a literal or a figurative sense. The answer does not mean that she is limited by Airs. Curran's vocabulary, but is an affirmative response to the question.

The word "put" in the preceding answer is one that requires some explanation, for it is frequently used by her, and makes some of her sayings difficult to understand. She makes it convey a number of meanings now obsolete, but it usually refers to her writings, her words, her sayings. She makes a noun of it, it will be noticed, as well as a verb. In the foregoing instance it means "tale," and it has a relation to the primary meaning of the verb, which is to place. The words that are put down become a "put," and the writer becomes a "putter." To a lady who told her that she had heard a sound like a bell in her ear, and asked if it was Patience trying to communicate with her, she answered dryly: "Think ye I be a tinkler o' brass? Nay. I be a putter o' words." Further to illustrate this use of the word, and also to throw an interesting light upon her method of communication and the reason for it, I present here a part of a conversation in which a Dr. Z. was the interrogator.

Dr. Z.—"Why isn't there some other means you could use more easy to manipulate than the Ouija board?"

Patience.—"The hand o' her (Mrs. Curran) do I to put (write) be the hand o' her, and 'tis ascribe (the act of writing) that setteth the one awhither by eyes-fulls she taketh in."

By this she seems to mean that if Airs. Curran tried to write for Patience with a pen or pencil, the act, being always associated with conscious thought, would set her consciousness to work, and put Patience "awhither."

Dr. Z.—"How did you know this avenue was open?"

Patience.—"I did to seek at crannies for to put; aye, and 'twer the her o' her who tireth past the her o' her, and slippeth to a naught o' putting; and 'twer the me o' me at seek, aye, and find. Aye, and 'twer so."

At the time Patience first presented herself to Mrs. Curran, she (Mrs. Curran) was very tired, and was sitting at the board with Mrs. Hutchings, with her head, as she expresses it, absolutely empty.

Dr. Z.—"Did you go forth to seek, or were you sent?" Patience.—"There be nay tracker o' path ne'er put thereon by sender."

Dr. Z.—"Did you know of the Ouija board and its use before?"

Patience.—"Nay, 'tis not the put o' me, the word hereon. 'Tis the put o' me at see o' her.

"I put athin the see o' her, aye and 'tis the see o' ye that be afulled o' the put o' me, and yet a put thou knowest not.

"That which ye know not o' thy day hath slipped it unto her, and thence unto thee. And thee knowest 'tis not the put o' her; aye, and thee knowest 'tis ne'er a putter o' thy day there be at such an put. Aye, and did he to put, 'twould be o' thy day and not the day o' me. And yet ye prate o' why and whence and where. I tell thee 'tis thee that knowest that which ye own not."

Dr. Z.—"Why don't we own it, Patience? Patience.—"'Tis at fear o' gab."

It is no easy task to untangle that putting of puts, but, briefly, it seems to mean that Patience does not put her words on the board direct, with the hands of Mrs. Curran, but transmits her words through the mind or inner vision of Mrs. Curran, and yet it is the word of Patience and not of Mrs. Curran that is recorded. This accords with Mrs. Curran's impressions. And thou knowest, Patience farther says, that it is not the language of her, and no writer of thy day would or could write in such a language as I make use of.

Returning to Dr. X. and his party. They were present again a few days after the interview just given, having with them a Miss J., a newspaper writer from an Ohio city. Dr. X. in the meantime had thought much upon the phenomena, and Patience immediately directed her guns upon the anatomist, in this manner:

Patience.—"Hark ye, lad, unto thee I do speak. Thou hast a sack o' the wares o' me, and thou hast eat therefrom. Yea, and thou hast spat that which thou did'st eat, and eat it o'er. And yet thou art not afulled.

"Hark! Here be a trick that shall best thee at thine own trick. Lo, thou lookest upon flesh and it be but flesh. Yea, thou lookest unto thy brother, and see but flesh. And yet thy brother speakest word, and thou sayest: 'Yea, this is a man, aye, the brother o' me.' Then doth death lay low thy brother, and he speak not word unto thee, thou sayest: 'Nay, this is no man; nay, this is but clay.' Then lookest thou unto thy brother, and thou seest not the him o' him. Thou knowest not the him o' him (the soul) but the flesh o' him only.

"More I tell thee. Thy very babe wert not flesh; yea, it were as dead afore the coming. Yet, at the mother's bearing, it setteth within the flesh. And thou knowest it and speak, yea, this is a man. And yet I tell thee thou knowest not e'en the him o' him! Then doth it die, 'tis nay man, thou sayest. Yet, at the dying and afore the bearing, 'twer what? The him o' him wert then, and now, and ever.

"Yea, I speak unto thee not through flesh, and thou sayest: This is no man, yea, for thine eyes see not flesh, yet thou knowest the me o' me, and I speak unto thee with the me o' me. And thou art where upon thy path o' learning!"

There was some discussion following this argument in which Dr. X. admitted that he accepted only material facts and believed but what he saw.

Patience.—"Man maketh temples that reach them unto the skies, and yet He fashioneth a gnat, and where be man's learning!

"The earth is full o' what the blind in-man seeth not. Ope thine eye, lad. Thou art athin dark, and yet drink ye ever o' the light."

Dr. X.—"That's all right, Patience, and a good argument; but tell me where the him o' him of my dog is."

Patience.—"Thou art ahungered for what be thine at the hand o' thee. Thy dog hath far more o' Him than thy brothers who set them as dogs and eat o' dog's eat. The One o' One, the All o' All, yea, all o' life holdeth the Him o' Him, thy Sire and mine! 'Tis the breath o' Him that pulses earth. Thou asketh where abides this thing. Aneath thy skull's arch there be nay room for the there or where o' this!

Miss J. then took the board and Patience said:

"She taketh it she standeth well athin the sight o' me that she weareth the frock o' me."

This caused a laugh, for it was then explained by the visitors that Miss J. had chosen to wear a frock somewhat on the Puritan order, having a gray cape with white cuffs and collar, and had said she thought Patience would approve of it.

Patience.—"Here be a one aheart ope, and she hath the in- man who she proddeth that be opeth his eyes. Yea, she seest that which be and thou seest not."

It was remarked that Patience was evidently trying to be very nice to Miss J.

Patience.—"Nay, here be a one who tickleth with quill, I did hear ye put. Think ye not a one who putteth as me, be not a love o' me? Yea, she be. And I tell thee a something that she will tell unto ye is true. Oft hath she sought for word that she might put, and lo, from whence she knoweth not it cometh."

Miss J. said this was true.

Patience.—"Shall I then sing unto thee, wench?" Miss J. expressed delight, and the song followed.

Ah, how do I to build me up my song for thee?
Yea, and tell unto thee of Him.
I'd shew unto thee His loving,
I'd shew unto thee His very face.
Do then to list to this my song.

Early hours, strip o' thy pure,
For 'tis the heart of Him.
Earth, breathe deep thy busom,
Yea, and rock the sea,
For 'tis the breath of Him.

Fields, burst ope thy sod,
And fling thee loose thy store,
For 'tis the robe of Him.

Skies, shed thou thy blue,
The depth of heaven,
For 'tis the eyes of Him.

Winter's white, stand thou thick
And shed thy soft o'er earth,
For 'tis the touch of Him.

Spring, shed thou thy loosened
Laughter of the streams,
For 'tis the voice of Him.

Noon's beat, and tire o' earth,
Shed thou of rest to His,
For 'tis the rest of Him.

Evil days of earth,
Stride thou on and smite,
For 'tis the frown of Him.

Earth, this, the chant o' me,
May end, as doth the works o' man,
But hark ye; Earth holdeth all
That hath been;
And Spring's ope, and sowing
O' the Winter's tide,
Shall bear the Summer's full
Of that that be no more.
For, at the waking o' the Spring,
The wraiths o' blooms agone
Shall rise them up from out the mould
And speak to thee of Him.
Thus, the songs o' me,
The works o' thee,
The Earth's own bloom,
Are HIM.

The interest of Dr. X. in this phenomenon brought an eminent psychologist, associated with one of the greatest state universities in the country, some distance from Missouri, for an interview with Patience. He shall be known here as Dr. V. With him and Dr. X. was Dr. K., a physician. Dr. V. sat at the board first, and Patience said to him:

"Here be a one, verily, that hath a sword. Aye, and he doth to wrap it o'er o' silks. Yea, but I do say unto thee, be doth set the cups o' measure at aright, and doth set not the word o' me as her ahere (Mrs. Curran). Nay, not till he hath seen and tasted o' the loaf o' me; and e'en athen he would to take o' the loaf and crumb o' it to bits and look unto the crumb and wag much afore he putteth. And he wilt be assured o' the truth afore the putting."

This was discussed as a character delineation.

Patience.—"I'd set at reasoning. Since the townsmen do fetch aforth for the seek o' me, and pry aneath the me o' me, then do thou alike. Yea, put thou unto me."

Dr. V.—"Why fear Death?"

Patience.—"Thou shouldst eat o' the loaf (her writings).

Ayea, 'tis right and meet that flesh shrinketh at the lash."

Dr. V. was told of her poems on the fear of death.

Dr. V.—"What do you think of the attempts to investigate you? Is it right?"

Patience.—"Ayea. And thou hast o' me the loaf o' the me o' me, and thou hast o' it afar more than thou hast o' thy brother o' earth, and yet they seek o' me and seek ever."

Dr. V.—"Have you ever lived?"

Patience.—"What! Think ye that I be a prater o' thy path and ne'er atrod? Then thou art afollied, for canst thou tell o' here?

Dr. V.—"When did you live on earth?"

Patience.—"A seed aplanted be watched for grow. Ayea, but the seed held athin the palm be but a seed, and Earth hath seeds not aplanted that she casteth forth, e'en as she would to cast forth me, do I not to cloak me much."

Dr. V.—"I understand; but can you not answer a little clearer the question I put?"

Patience.—"The time be not ariped for the put o' this."

Dr. V.—"What does Lethe mean?

Patience.—"This be a tracker! Ayea, 'tis nay a word o' thy day or yet the word o' thy brother, that meaneth unto me. I be a maker o' loaf for the hungered. Eat thou. 'Tis not aright that thou shouldst set unto the feast athout thou art fed."

By this she seemed to mean that she wanted him to read her writings and see what it is she is endeavoring to do. She continued:

"Brother, this be not a trapping o' thy sword, the seeking o' me. Nay, 'tis ahind a cloak I do for to stand, that this word abe, and not me."

Mr. Curran here stated that this had ever been so; that Patience had obscured herself so that her message could not be clouded.

Patience.—"Aright. I do sing.
Gone! Gone! Ayea, thou art gone!
Gone, and earth doth stand it stark.
Gone! Gone! The even's breath Doth breathe it unto me
In echo soft; yea, but sharped, And cutting o' this heart.
Gone! Gone! Aye, thou art gone!
The day is darked, and sun
Hath sorried sore and wrapped him in the dark.
Gone! Gone! This heart doth drip o' drops
With sorry singing o' this song.
Gone! Gone! Yea, thou art gone!
And where, beloved, where?
Doth yonder golden shaft o' light
That pierceth o' the cloud
Then speak unto this heart?
Art thou athin the day's dark hours?
Hast thou then bid from sight o' me,
And yet do know mine hour?
Gone! Gone! What then hath Earth?
What then doth day to bring
To this the sorry-laden heart o' me,
That weepeth blood drops here?
Gone! Gone! Yea, but bark!
For I did trick the sorry, loved;
For where e'er thou art am I.

Yea, this love o' me shall follow thee
Unto the Where, and thou shalt ever know
That though this sorry setteth me
I be where'er thou art."

After this Dr. K., who resides in St. Louis, took the board.

Patience.—"Here abe a townsman. Aye, a Sirrah who knoweth men and atruth doth ne'er acloak the blade o' him as doth brother ayonder. Ayea, ahind a chuckle beeth fires.

"There abe weave 'Pon the cloth o' me, yea, but 'tis nay ariped the time that I do weave. Yea, thou hast a pack o' tricks. Show unto me, then, thine."

Here Dr. V. asked: "Do you know Dr. James?"

This referred to the late Dr. William James, the celebrated psychologist of Harvard.

Patience.—"I telled a one o' the brothers and the neighbors o' thy day, and he doth know."

She had given such an answer to a frequent visitor who bad inquired as to her knowledge of several eminent men long since dead. It was considered an affirmative answer.

Dr. V.—"Have you associated with Dr. James?"

Patience.—"Hark! Unto thee I do say athis; 'tis the day's break and Earth shall know, e'en athin thy day, much o' the Here.

"This, the brother o' ye, the seeker o' the Here, hath set a promise so, and 'tis for to be, I say unto thee. Thou knowest 'tis the word o' him spaked in loving. Yea, for such a man as the man o' him wert, standeth as a beacon unto the Here."

Dr. V.—"Could Dr. James, by seeking as you did, communicate with someone here as you are doing?"

Patience.—"This abe so; he who seeketh abe alike unto thee and thee. Ayea, thee and thy brother do set forth with quill, and thou dost set aslant, and with thy hand at the right o' thee. And thy brother doth trace with the hand at the left of him. And 'tis so, thou puttest not as him. This, the quill o' me, be for the put o' me, and doth he seek and know the trick o' tricks o' sending out a music with the quill o' me, it might then be so."

This was interpreted as meaning that if Dr. James could find one who had the conditions surrounding Mrs. Curran, and was able to master the rhythm which Patience uses to give the matter to her, then he could do it.

When the record of the foregoing interview was being copied, Mrs. Curran felt an impulse to write. Taking the board, Patience indicated that she had called, and at once set forth, apparently for Dr. V., the following explanation of her method of communication and the principle upon which it is based:

Patience.—"Aye, 'tis a tickle I be. Hark, there be a pulse— Nay, she (Mrs. Curran) putteth o' the word! Alist.—There abe a throb; yea, the songs o' Earth each do throb them, like unto the throbbing o' the heart that beareth them. Yea, and there be a kinsman o' the heart that beareth them. Yea, and there be a kinsman o' thee who throbbeth as dost thou. Yea, and be knoweth thee as doth nay brother o' thee whose throb be not as thine. So 'tis, the drop that falleth athin the sea, doth sound out a silvered note that no man heareth. Yet its brother drops and the drop o' it do to make o' the sea's voice. Aye, and the throb o' the sea be the throb o' it. So, doth thy brother seek out that he make word unto thee from the Here, he then falleth aweary. For thee of Earth do hark not unto the throb. And be the one aseeked not attuned unto the throb o' him he findeth, 'tis nay music. So 'tis, what be the throb o' me and the throb o' her ahere, be nay a throb o' music's weave for him aseek.

"I tell thee more. The throb hath come unto thy day long and long. Yea, they be afulled o' throb, and yet nay man taketh up the throbbing as doth the sea. The drop o' me did seek and find, and throb met throb o' loving. Yea, and even as doth the sea to throb out the silvered note o' drop, even so doth she to throb out the love o' me."

This seems, in effect, a declaration that communications of this character are a matter of attunement, possible only between two natures of identical vibrations, one seeking and the other receptive. It indicates too that her rhythmical speech has an influence upon the facility of her utterances. At another time she described her own seeking in this verse:

How have I sought!
Yea, how have I asought,
And seeked me ever through the earth's hours, Amid the damp, cool moon, when winged scrape Doth sound and cry unto the day

The waking o' the hosts!
Yea, and 'mid the noon's beat,
When Earth doth wither 'neath the sun,
And rose doth droop from sun's-kiss,
That stole the dew; and 'mid the wastes
O' water where they whirl and rage,
And seeked o' word that I
Might put to answer thee.
Ayea, from days have I then stripped
The fulness of their joys, and pryed
The very buds that they might ope for thee.
Aye, and sought the days apast,
That I might sing them unto thee.
And ever, ever, cometh unto me
Thy song o' why? why? why?
And then, lo, I found athin this heart
The answer to thy song.
Aye, it chanteth sweet unto this ear,
And filleth up the song.
Do hark thee, bark unto the song,
For answer to thy why? why? why?
I sing me Give! Give! Give!
Aye, ever Give!

When the foregoing verse was received, Dr. X. was again present, this time with his wife and two physicians, Dr. R. and Dr. P. It will have been observed that many doctors of many kinds have "sat at the feet" of Patience Worth, but all, as I have said, have come as the friends of friends of Mrs. Curran, upon her invitation, or upon that of Mr. Curran. On this occasion Patience began:

"They, do seek o' me, ever; that they do see the pettiskirt o' me, and eat not o' the loaf! '(More interested in the phenomenon than the words.) Ayea, but he ahere (Dr. R.) hath a wise pate. Aye, he seeketh, and deep athin the heart o' him sinketh seed o' the word o' me. Aye, even though he doth see the me o' me athrough the sage's eye o' him, still shall he to love the word o' me."

After due acknowledgments from Dr. R., she continued:

"Yea, brother, hark unto the word o' me, for thou dost seek amid the fields o' Him! Aye, and 'tis, thou knowest, earth's men that be afar amore awry athin the in-man than in the flesh. And 'tis the in-man o' men thou knowest."

Dr. R., a neurologist, gave hearty assent.

"Put thou unto me. (Question me.) 'Tis awish I be that ye weave."

Dr. R.—"Do you see through Mrs. Curran's eyes and hear through her ears?"

Patience.—"Even as thou hast spoke, it be. Aye, and yet I say me 'tis the me o' me that knoweth much she heareth and seeth not."

Then to a question had she ever talked before with anyone, she said: "Anaught save the flesh o' me."

"Fetch ye the wheel," she commanded, "that I do sit and spin."

This was one of her ways of saying that she desired to write on her story, and she dictated several hundred words of it, after which Dr. P. took the board and she said:

"What abe ahere? A one who seeth sorry and maketh merry! Yea, a one who leaveth the right hand o' him unto its task, and setteth his left at doing awry o' the task o' its brothers. Aye, he doeth the labors o' his brother, aye, and him. Do then, aweave."

In compliance some more of the story was written, and then Dr. R. "wondered" why he could not write for Patience, to which she answered:

"Hark unto me, thou aside. Thou shalt put (say) 'tis her ahere (i.e., Mrs. Curran, who does it); ayea, and say much o' word, and e'en set down athin thy heart thy word o' what I be, and yet I tell thee, I be me! Aye, ever, and the word o' me shall stand, e'en when thou and thou art ne'er ahere!

"E'en he who doth know not o' the Here hath felt the tickle o' my word, and seeketh much this hearth.

"Then eat thee well and fill thee up, and drink not o' the brew o' me and spat forth the sup. Nay, fill up thy paunch. 'Twill merry thee!"

Dr. P. asked her a question about her looks.

"'Tis a piddle he putteth," she said.

And now we come to a sitting of a lighter character. There were present at this Dr. and Mrs. D., Mr. and Mrs. M. and Mrs. and Miss G.

"Aflurry I be!" cried Patience. "Aye, for the pack o' me be afulled o' song and weave, and e'en word to them ahere.

"Yea, but afirst there be a weave, for the thrift-bite eateth o' me." (The bite of her thrifty nature.)

Some of the story followed and then she said to Mrs. M., who sat at the board:

"Here be aone who doth to lift up the lid o' the brew's pot, that she see athin! Aye, Dame, there abe but sweets athin the brew for thee. Amore, for e'en tho' I do brew o' sweets and tell unto thee, I be a dealer o' sours do I to choose! Ayea, and did I to put the spatting o' thee athin the brew, aye verily 'twould be asoured a bit!" Then deprecatingly: "'Tis a piddle I put!

"Yea, for him aside who sitteth that he drink o' this brew do I to sing; fetch thee aside, thee the trickster o' thy day!"

There being so many "tricksters" in the room, they were at a loss to know which one she meant. Mr. C. asked if she meant Dr. D., but Patience said:

"Thinkest thou he who setteth astraigbt the wry doth piddle o' a song? Anay, to him who musics do I to sing."

This referred to Mr. G., who is a musician and a composer, and be took the board. Patience at once gave him this song:

Nodding, nodding, 'pon thy stem,
Thou bloom o' morn,

Nodding, nodding to the bees,
Asearch o' honey's sweet.
Wilt thou to droop and wilt the dance o' thee,
To vanish with the going o' the day?
Hath the tearing o' the air o' thy sharped thorn
Sent musics up unto the bright,
Or doth thy dance to mean anaught
Save breeze-kiss 'pon thy bloom?

Hath yonder songster barked to thee,
And doth be sing thy love?
Or hath be tuned his song of world's wailing o' the day?
Doth mom shew thee naught save thy garden's wall
That shutteth thee away, a treasure o' thy day?
Doth yonder bum then spell anaught,
Save whirring o' the wing that hovereth
O'er thy bud to sup the sweet?

Ah, garden's deep, afulled o' fairies' word,
And creeped o'er with winged mites,
Where but the raindrops' patter telleth thee His love—
Doth all this vanish then, at closing o' the day?
Anay. For He hath made a one who seeketh here,
And storeth drops, and song, and bum, and sweets,
And of these weaveth garland for the earth.
From off his lute doth drip the day of Him.

Patience then turned her attention to Mr. M., saying:

"Ayea, he standeth afar from the feasting place and doth to smack him much!"

Mr. M. took the board, and she began to talk to him in an intimate way about the varying attitudes of people toward her and her work, and what they say of her:

"I be a dame atruth," she said, "and I tell thee the word o' wag that shall set thy day, meaneth anaught but merry to me. Hark! I put a murmur o' thy day, for at the supping o' this cup the earth shall murmur so:

"'Tis but the chatter o' a wag! Aye, the putting o' the mad! 'Tis piddle! Yea, the trapping o' a fool! Yea, 'tis but the dreaming o' the waked! Aye, the word o' a wicked sprite! Yea, and telleth naught and putteth naught!

And yet, do harken unto me. They then shall seek to taste the brew and sniff the whiffing o' the scent; ayea, and stop alonger that they feast! And lo, 'twill set some asoured, and some asweet; aye and some, ato (too), shall fill them upon the words THEY do to put o' me, and find them filled o' their own put, and lack the room for cat o' the loaf o' me. 'Tis piddle, then! Aye, and yet I say me so, 'tis bread, and bread be eat though it be but sparrows that do seek the crumb. Then what care ye? For bake asurely shall be eat!"

This is a point she often makes, and strives earnestly to impress—that whatever she may be, whatever the world may think she is, there is substance in her words. It is bread, and will be eaten, if only by the sparrows. So, she is content. She has put this thought, somewhat pathetically, into the little verse which follows:

Loth as Night to dark o' Day,
Loth do I to sing.

Aye, but doth the Day aneed a song,
'Tis they, o' Him,
The songsters o' the Earth,
Do sing them on, to Him.
What though 'tis asmiled?
And what Though 'tis nay aseeked o' such a song?
Aye, what though 'tis sung 'mid dark?

'Tis I would sing,
Do thee to list, or Day.

I be a dame who knoweth o' the hearth. Aye, and do to know o' the hearts o' men," she said to Mrs. D., who next took the place with Mrs. Curran. "Ayea, and do to put o' that athin the hearts o' them that doth tickle o' their merry! This be a tale for her ahere."

The Story of the Herbs

Lo, there wert a dame and her neighbor's dame and her neighbor's dame. And they did to plant them o' their gardens full. And lo, at a day did come unto the garden's ope a stranger, who bore him of a bloom-topped herb. And lo, be spaked unto the dame who stood athin the sun-niche that lay at the garden's end, and he did tell unto her of the herb he bore. And lo, be told that be would give unto her one of these, and to her neighbor dame a one, atoo (also), and to her neighbor dame a one atoo, and he then would leave the garden's place and come at the fulling o' the season-tide when winter's bite did sear, and that he then would seek them out, and they should shew unto him the fulling o' the herb.

"And lo, be went him out unto the neighbor's dame and telled unto her the same, and to her neighbor's dame the same, and they did seek one the other and tell o' all the stranger bad told unto them. And each had sorry, for feared 'twer the cunger o' the wise men, and each aspoke her that she would to care and care for this the herb he did to leave, and that she would have at the fulling o' the season the herb that stood at the fullest bloom. And each o' the dames did speak it that this herb o' her should be the one waxed stronger at the fulling. And lo, none told unto the other o' how this would to be.

"And lo, the first o' dames did plant her herb adeep and speak little, and lo, her neighbor dames did word much o' the planting, and carried drops from out the well that the herbs might full. And lo, they did pluck o' the first bud that them that did follow should be afuller. And lo, the dame afirst o' the garden the stranger did to seek, did look with sunked heart at the thriving o' the herbs o' the neighbor dames. And lo, she wept thereon, and 'twer that her well did dry, and yet she seeked not the wells of her sisters. Nay, but did weep upon the earth about the herb, and lo, it did to spring it up. And lo, she looked not with greed upon her sister's herb; nay, for at the caring for the bloom, lo, she loved its bud and wept that she had nay drop to give as drink unto it.

"And lo, at a certain day the stranger came and did seek the dames, and came him unto her garden where the herb did stand, and be bore the herbs of her sisters, and they wert tall and full grown and filled o' bloom. And be did to put the herb o' her sisters anext the herb o' her, and lo, the herb o' her did spring it up, and them o' her sisters shrunked to but a twig. And he did call unto the dames and spake:

"'Lo, have ye but fed thy herb that it be full o' bloom, that thou shouldst glad thee o'er thy sister? And lo, the herb o' her hath drunked her tears shed o' loving, and standeth sweet- bloomed from out the tears o' her.'

"And lo, the herb did flower aneath their very eyes. And lo, the flowering wert fulled o' dews-gleam, and 'twer the sweet o' her heart, yea, the dew o' heaven."

Following this pretty parable someone spoke of a newspaper article that had appeared that day, and Patience remarked:

"'Tis a gab o' fool. Aye, and the gab o' fool be like unto a spring that be o'erfull o' drops, 'tis ne'er atelling when it breaketh out its bounds."

With this sage observation she dismissed the fool "as unworthy of further consideration, and gave this poem:

Do I to love the morn,
When Earth awakes, and streams
Aglint o' sun's first gold,
As siren's tresses thred them through the fields;
When sky-cup gleameth as a pearl;
When sky-hosts wake, and leaf bowers
Wave aheavied with the dew?

Do I to love the eve,
When white the moon doth show,
And frost's sweet sister, young night's breath,
Doth stand aglistened 'pon the blades;
When dark the shadow deepeth,
Like to the days agone that stand
As wraiths adraped o' black
Along the garden's path;
When sweet the nestlings twitter
'Neath the wing of soft and down
That hovereth it there within
The shadows deep atop the tree?

Do I to love the mid-hours deep—
The royal color o' the night?
For earth doth drape her purpled,
And jeweled o'er athin this hour.

Do I to love these hours, then,
As the loved o' me?
Nay, for at the morn,
Lo, do I to love the eve!
And at the eve,
Lo, do I to love the mom!
And at the morn and eve,
'Tis night that claimeth me.

A little of the reasoning of Patience upon Earth questions may appropriately come in here. The Currans, with a single visitor, had talked at luncheon of various things, beginning with music and ending with capital punishment, the latter suggested by an execution which at the moment was attracting national attention. When they took the board, after luncheon, Patience said:

"List thee. Earth sendeth up much note. Yea, and some do sound them at wry o' melody, and others sing them true. And lo, they who sing awry shall mingle much and drown in melody. And I tell thee, o'er and above shall sound the note o' me!"

And then she gave them to understand that she had listened to their discussion!

"Ye spake ye of eye for eye. Yea, and tooth for tooth. Yea, but be thy brother's eye not the ope o' thine, then 'tis a measure less the full thou hast at taking o' the eye o' him. Yea, and should the tooth o' him put crave for carrion, and thine for sweets, then bow doth the tooth o' him serve thee?"

Here the sitters asked: "How about a life for a life, Patience?"

Patience.—"Ye fill thy measure full o' sands that trickle waste at each and every putting. I tell thee thou hast claimed life; aye, and life be not thine or yet thy brother's for the taking or giving. Yea, and such an soul hath purged at the taking or giving, and rises to smile at thy folly.

"Aye, and more. List! The earth's baggage, hate, and might, and scorn, fall at earth's leave, a dust o' naught, like the dust o' thy body crumbleth.

"Thou canst strip the body, yea, but the soul defieth thee!"

The visitor referred to in the preceding talk is a frequent guest of the Currans, and is one of the loved ones of Patience. This visitor, who is a widow, remarked one evening that Patience was deep and lived in a deep place.

"Aye," said Patience, "a deeper than word. There be ahere what thou knowest abetter far than word o' me might tell. (This seems to refer to the visitor's husband.) Ayea thou hungereth, and bread be thine, for from off lips that spaked not o' the land o' here in word o' little weight, thou hast supped of love, and know the path that be atrod by him shall be atrod even so by thee, e'en tho' thou shouldst find the mountain's height and pits o' depth past Earth's tung.

"Shouldst thou at come o' here to hark unto the sound of this voice, thinkest thou that heights, aye or depths, might keep thee from there? And even so, doth not the one thou seeketh too, haste e'en now to find the path and waiteth?

"Then thinkest thou this journey be lone? Nay, I tell thee, thou art areach e'en past the ye o' ye, and he areach ato. Then shall the path's ope be its end and beginning. In love is the end and beginning of things.

"Yea, yea, yea, the earth suppeth o' the word o' me, and e'en at the supping stoppeth and speaketh so. What that one not o' me doth brew. Thou knowest this, dame. Aye, but what then? And why doth not the blood o' me speak unto me?

"'Tis a merry I be. Lo, have I not fetched forth unto a day that holdeth little o' the blood o' me, that I might deal alike unto my brother and bring forth word that be ahungered for aye, and they speak them o' her ahere and wag and hark not? Yea, and did the blood o' them spake out unto their very ears I vow me 'twould set the earth ariot o' fearing. Yea, man loveth blood that hath not flowed, but sicketh o'er spilled blood. Yea, then weave."

There was some discussion following this, to the effect that whatever explanations might be given of this phenomenon, many would believe in Patience Worth as an independent personality, which brought from her the following discourse which may well conclude these conversations:

"Yea, the tooth o' him who eateth up the flesh I did to cloak me athin, shall rot and he shalt wither. Aye, and the word o' me shalt stand. Fires but bake awell.

"Sweet hath the sound of the word o' Him asounded unto the ears o' Earth that hark not.

"Yea, and He hath beat upon the busom of Earth and sounded out a loud noise, and Earth harkened not.

"And He hath sung thro' the mother's songs o' Earth, and Earth harkened not.

"Yea, and He hath sent His own with word, and Earth harkened not.

"Then 'tis Earth's own folly that batheth her.

"Yea, and Folly cometh astreaming ribbands, and showering color, and grinning 'pon his way.

"Yea, but Folly masketh and leadeth Earth and man assuredly unto Follies pit—self. And self is blind.

"Then whence doth Earth to turn for aid? For Folly followeth not the blind, and the voice of him who falleth unto the pit of Folly soundeth out a loud note. Yea, and it echoeth self.'

"And lo, the Earth filled up o' self, hearketh not unto the words of Him, the King of Wisdom.

Yea, and I say unto thee, though them o' Him fall pierced and rent athin the flow o' their own blood thro' the self-song o' his brother, he doeth this for Him.

"And the measuring rod shall weight out for him who packeth the least o' self athin him, afull o' measure, and light for him who packeth heavy o' self.

"Ayea, and more. I speak me o' lands wherein the high estate be self. Yea, yea, yea, o' thy lands do I to speak. Woe unto him who feareth that might shall slay! Self may wield a mighty blow, but it slayeth never.

"'Tis as the dame who watcheth o'er her brood, and lo, this one hath sorry, and that one hath sorry. And she flitteth here and yon, and lo, afore she hath fetched out the herbs, they sleep them peaceful. So shall it be at this time. The herbs shall be fetched forth but lo, the lands shall sleep them peaceful.

"Yea, for Folly leadeth, and Wisdom warreth Folly."

Chapter 7

RELIGION

Teach me that I be Ye.

AND now we well may ask: What is the purpose of all this? Here we appear to have an invisible intelligence, speaking an obsolete language, producing volumes of poetry containing many evidences of profound wisdom. So far as I have been able to find out, no such phenomenon has occurred before since the world began. Do not misunderstand that assertion. There is nothing extraordinary in the manner of its coming, as I have said before. The publications of the Society for Psychical Research are filled with examples of communications received in the same or a similar way. The fact that makes this phenomenon stand out, that altogether isolates it from everything else of an occult nature, is the character and quality of its literature. Literature is some thing tangible, something that one can lay hands on, so to speak. It is in a sense physical; it can be seen with the eyes. And this literature is the physical evidence which Patience Worth presents of herself as a separate and distinct personality.

But why is it contributed? Is there in it any intimation or assertion of a definite purpose?

If we may assume that Patience is what she seems to be— a voice from another world, then indeed we may discern a purpose. She has a message to deliver, and she gives the impression that she is a messenger.

"Do eat that which I offer thee," she says. "'Tis o' Him. I but bear the pack apacked for the carry o' me by Him."

Constantly she speaks of herself as bearing food or drink in her words. "I bid thee eat," she said to one, "and rest ye, and eat amore, for 'tis the wish o' me that ye be filled." The seed, the loaf, the cup, are frequently used symbolically when referring to her communications.

There be a man who buyeth grain and he telleth his neighbor and his neighbor's neighbor, and lo, they come asacked and clamor for the grain. And what think ye? Some do make price, and yet others bring naught. But I be atelling ye, 'tis not a price I beg. Nay, 'tis that ye drink my cup."

"'Tis truth o' earth that 'tis the seed aplanted deep that doth cause the harvester for to watch. For lo, doth he to hold the seed athin (within) his band, 'tis but a seed. And aplanted he doth watch him in wondering. Verily do I say, 'tis so with me. I be aplanted deep; do thee then to watch."

And with greater significance she has exclaimed:

"Morn hath broke, and ye be the first to see her light. Look ye wide-eyed at His workings. He hath offered ye a cup."

It is thus she announces herself to be a herald of a new day, a bearer of tidings divinely commissioned.

What, then, is her message? For answer it may be said that it is at once a revelation, a religion and a promise. Whatever we may think of the nature of this phenomenon, Patience herself is a revelation, and there are many revelations in her words. The religion she presents is not a new one. It is as old as that given to the world nineteen centuries ago; for fundamentally it is the same. It is that religion, stripped of all the doctrines and creeds and ceremonials and observances that have grown up about it in all the ages since His coming, and paring it down to the point where it can be expressed by the one word— Love. Love, going out to fellow man, to all nature and overflowing toward God.

In the consideration of this religion let us begin at the beginning, at the ground, so to speak, with this expression of love for the loveless:

Ah, could I love thee,
Thou, the loveless o' the earth,
And pry aneath the crannies
Yet untouched by mortal band

To send therein this love o' mine
Thou creeping mite, and winged speck,

And whirled waters o' the mid o' sea
Where no man seeth thee?
And could I love thee, the days
Unsunned and laden with bate o' sorrying?
Ah, could I love thee,
Thou who beareth blight;
And thou the fruit bescorched
And shrivelling, to fall unheeded
'Neath thy mother-stalk?

Ah, could I love thee, love thee?
Aye, for Him who loveth thee,
And blightest but through loving;
Like to him who bendeth low the forest's king
To fashion out a mast.

Love for everything is the essence of her thought and of her song. And as she thus sings for the loveless, so she sings for the wearied ones and the failures of the earth:

I'd sing ,
Wearied word adropped by weary ones,
And broked mold afashioned out by wearied bands;
A falter-song sung through tears o' wearied one;
fancied put o' earth's fair scene
Unfinished, aye, o'ertaken by sore weariness
O' thee I'd sing.

Aye, and put me such an songed-note
That earth, aye, and heaven, should hear;
And thou, aye all o' ye, the soul-songs
O' my brothers, be afinished,
At the closing o' my song.

Aye, and wearied, aye and wearied,
I'd sing. I'd sing for them, the loved o' Him,
And brothers o' thee and me. Amen.

This is the prelude and now comes the song:

I choose o' the spill
O' love and word and work,
The waste o' earth, to build.

Ye bark unto the sages,
And oft a way-singer's song
Hath laden o'erfull o' truth,
And wasteth 'pon the air,
And falleth not unto thine ear.

Think ye He scattereth whither
E'en such an grain? Nay.
And do ye seek o' spill

And put unto thy song, '
Twill fill its emptiness.

Ye seek to sing but o' thy song,
And 'tis an empty strain. 'Tis need
O' love's spill for to fill.

The spill of earth, the love that goes unnoticed and unappreciated, the words that are unheard or unheeded, the work that seems to be for naught—none of these is waste. A song it is for the wearied ones, the heart-sick and discouraged, "the loved of Him and brothers of thee and me."

And yet she calls them waste but to show that they are not. "The waste of earth," she says, "doth build the Heaven," and this is the theme of much of her song.

Earth hath filled it up o' waste and waste.
The sea's fair breast, that heaveth as a mother's,
Beareth waste o' wrecks and wind-blown waste.
The day doth hold o' waste.
The smiles that die, that long to break,
The woes that burden them already broke,
'Tis waste, ah yea, 'tis waste.

And yet, and yet, at some fair day,
E'en as the singing thou dost note
Doth bound from yonder bill's side green
As echo, yea, the ghost o' thy voice;
So shall all o' this to sound aback
Unto the day.
Of waste, of waste, is heaven builded up.

It is to the waste of earth that she speaks in this message of love and sympathy:

Ah, emptied heart! The weary o' the path!
How would I to fill ye up o' love!
I'd tear this lute, that it might whirr
A song that soothed thy lone, awearied path.
I'd steal the sun's pale gold,
And e'en the silvered even's ray,
To treasure them within this song
That it be rich for thee.
From out the wastes o' earth I'd seek
And catch the woe-tears shed,
That I might drink them from the cup
And fill it up with loving.
From out the hearts afulled o' love
Would I to steal the o'er-drip
And pack the emptied hearts of earth.
The bread o' love would I to cast
Unto thy bywayed path, and pluck me
From the thorned bush that traileth o'er
The stepping-place, the thorn, that brothers
O' the flesh o' me might step 'pon path acleared.
Yea, I'd coax the songsters o' the earth
To carol thee upon thy ways,
And fill ye up o' love and love and love.

And a message of cheer and encouragement she gives to those who sorrow, in this:

"The web o' sorrow weaveth 'bout the days o' earth, and 'tis but Folly who plyeth o' the bobbin. I tell thee more, the bobbins stick and threads o' day-weave go awry. But list ye; 'tis he who windeth o' his web 'pon smiles and shuttleth 'twixt smiles and woe who weaveth o' a day afull and pleantious. And sorrow then wilt rift and show a light athrough."

Smiles amid sorrows. He who windeth of his web upon smiles not only rifts his own woes but those of others, as she expresses it in this verse:

The smile thou cast today that passed
Unheeded by the world; the handclasp
Of a friend, the touch of baby palms
Upon its mother's breast—
Whither have they flown along the dreary way?
Mayhap thy smile
Hath fallen upon a daisy's golden head,
To shine upon some weary traveler
Along the dusty road, and cause
A softening of the hard, hard way.
Perchance the handclasp strengthened wavering love
And lodged thee in thy friend's regard.
And where the dimpled hands caress,
Will not a well of love spring forth?
Who knows, but who will tell
The hiding of these fleeting gifts!

And she gives measure to the same thought in this:

Waft Ye through the world sunlight;
Throw ye to the sparrows grain
That runneth o'er the heaping measure.
Scatter flower petals, like the wings
Of fluttering butterflies, to streak
The dove-gray day with daisy gold,
And turn the silver mist to fleece of gold.
Hath the king a noble who is such
An wonder-worker? Or hath his jester
Such a pack of tricks as thine?

Both of these last have to do with the hands and with the use of the hands in the expression of love for others, but in the following poem

Patience pays a tender and yet somewhat mystical tribute to the hands themselves, empty hands filled with the gifts of Him, the power to build and weave and soothe:

Hands. Hands. The hands o' Earth;
Abusied at fashioning, Aye,
And put o' this, aye, and that.
Hands. Hands upturned at empty.
Hands. Hands untooled, aye, but builders
O' the soothe o' Earth.

Hands. Hands aspread, aye, and sending forth
That which they do hold—the emptiness.
Aye, at empty they be, afulled o' the give o' Him.
At put at up, aye, and down, 'tis at weave
O' cloth o' Him they be.

Hands. Hands afulled o' work o' Him;
'Aye, and ever at a spread o' doing in His name.
'Aye, and at put o' weave
For naught but loving.

There are no doubt such hands on earth, many of them "ever at a spread of doing in His name," but not often have their work and their mission been so beautifully and so fittingly expressed as in this strange verse which, to me at least, grows in wonder at every reading. And this not so much because of the quaintness of the words and the singularity of the construction, as for the thought. This, however, is characteristic of all of her work. There is always more in it than appears upon the surface. And yet when one analyzes it, one finds that whatever may be the nature or the subject of the composition, in nearly every instance love is the inspiration.

The love that she expresses is universal. It goes out to nature in all its forms, animate and inanimate, lovely and unlovely. It is manifested in all her references to humanity, from the infant to doddering age; and her compositions are filled with appeals for the application of love to the relations between man and man. But it is when she sings of God that she expresses love with the most tender and passionate fervency—His love for man, her love for Him. "For He knoweth no beginning, no ending to loving," she says, "and loveth thee and me and me and thee ever and afore ever." "Sighing but bringeth up heart's weary; tears but wash the days acleansed; bands abusied for them not thine do work for Him; prayers that fall 'pon but the air and naught, ye deem, sing straight unto Him. Close, close doth He to cradle His own to Him." She gives poetic expression to this divine love in the song which follows:

Brother, weary o' the plod,
Art sorried sore o' waiting?
Brother, bowed aneath the pack o'
Earth, Art seeking o' the path
That leadest thee unto new fields
O' green, and breeze-kissed airs?
Art bowed and bent o' weight o' sorry?
Art weary, weary, sore?
Then come and bark unto this song o' Him.

Hast thou atrodden 'pon the Earth,
And worn the paths o' folly
Till thou art foot-sore?
And hast the day grinned back to thee,
A folly-mask adown thy path
That layeth far behind thee?
Thy heart, my brother, hast thou then
Alost it 'pon the path?
And filled thee up o' word and tung
O' follysingers long the way?

Ah, weary me, ah, weary me!
Come thou unto this breast.
For though thou hast suffered o' the Earth,
And though thy robe be stained
O' travel o'er the stoney way,
And though thy lips deny thy heart,
Come thou unto this breast,
The breast o' Him.
For He knoweth not the stain.
Aye, and the land o' Him doth know
No stranger 'mid its hosts.
Ayea, and though thou comest mute,

This silence speaketh then to Him,
And He doth bold Him ope His arms.

So come thou brother, weary one,
To Him, for 'tis but Earth and men
Who ask thee WHY.

She pours out her love for God in many verses of praise and prayer.

Bird skimming to the south,
Bear thou my song,
Sand slipping to the wave's embrace,
Do thou but bear it too!
And, shifting tide, take thou
Unto thy varied paths
The voicing of my soul!
I'd build me such an endless
Chant to sing of Him
That days to follow days
Would be but builded chord
Of this my lay.

Still more ardently does she express her love in these lines:

Spring, thou art but His smile
Of happiness in me, and sullen days
Of weariness shall fall when Spring is born
In winds of March and rains of April's tears.
Methinks 'tis weariness of His that I,
His loved, should tarry o'er the task
And leave life's golden sheaves unbound.
And, Night, thou too art mine, of Him.
Thy dim and veiled stars are but the eyes
Of Him that through the curtained mystery
Watch on and sever dark from me.
And, Love, thou too art His,
His words of wooing to my soul.
Should I, then, crush thee in embrace,
And bruise thee with my kiss,

And drink thy soul through mine?
What, then! 'Tis He, 'tis He, my love,
That gave me thee, and while my love is thine,
What wonder is it causeth here
This heart of mine to stifle so
And seek expression in a prayer of thanks?

With equal fervency of devotion and gratitude she sings this tribute to the day:

Ah, what a day He hath made, He hath made!
It flasheth abright and asweet, and asweet.
It showeth His love and His smile, yea, His smile.

The bills stand abrown, aye astand brown,
And peaked as a monk in his cowl, aye, his cowl!
The grass it hath seared, aye, hath seared
And scenteth asweet, yea, asweet.

Ayonder a swallow doth whirl, aye, doth whirl,
And skim mid the grey o' the blue,
Aye, the grey o' the blue.
The young wave doth lap 'pon the sands,
Yea, lap soft and soft 'pon the sands.
The field's maid doth seek, yea, doth seek,
And send out her song to the day,
Yea, send out her song to the day.

My heart it is full, yea, 'tis full,
For the love of Him batheth the day,
Yea, the love of Him batheth the day.

Ah, what a day He hath made,
Yea, He hath made it for me!

Her prayers are not appeals for aid; they are not begging petitions. They are outpourings of love and trust and gratitude.

To an old couple, friends of Mr. and Mrs. Curran, who passed a round-eyed evening with Patience, she said:

Keep ye within thy heart a song And murmur thou this prayer:

"My God, am I then afraid
Of heights or depths?
And doth this dark benumb my quaking limbs?
And do I stop my song in fear
Lest Thee do then forsake me?
Nay, for I do love Thee so,
I fain would choose a song
Built from my chosen tung,
And though it be but chattering
Of a soul bereft of reasoning,
I know Thou would'st love it as Thine own,
For I do love Thee so!"

This was not given for another, but is her own cry:

I beseech Thee, Lord, for naught!
But cry aloud unto the sunlight
Who bathes the earth in gold
And boldly breaketh into crannies
Yet unseen by man:
Flash thou in flaming sheen!
Mine own song of love doth falter
And my throat, it is afail!

And thou, the greening shrub along the way,
And earth at bud-season,
Do thou then spurt thy shoots
And pierce the air with loving!

And age-wabbled brother—
I do love thee for thy spending,
And I do gaze in loving at thy face,
Whereon I find His peace,
And trace the withered cheek
For record of His love.
Around thy lips doth hang
The child-smile of a trusting heart;
And world hath vanished

From thine eyes, bedimmed
To gard thee at awakening.
Thou, too, art of my song of love.

I beseech Thee, Lord, for naught.
These bands are Thine for loving,
And this heart, already Thine,
Why offer it?
I beseech Thee, Lord, for naught.

This one does ask for something, but only to know Him:

Teach me, O God,
To say', "'Tis not enough."
Aye, teach me, O Brother,
To sing, and though the weight
Be past this strength,
Teach me, O God, to say,
"'Tis not enough—to pay!'"

Teach me, O God, for I be weak.
Teach me to learn
Of strength from Thee.

Teach me, O God, to trust, and do.
Teach me, O God, no word to pray.
Teach me, O God, the heart Thou gavest me.
Teach me, O God, to read thereon.
Teach me, O God, to waste not word.
Teach me that I be Ye!

That last line presents the most impressive principle of the religion she expresses, and which, we might almost say, she embodies.

"Who are you?" she was once asked abruptly.

"I be Him," she replied; "alike to thee. Ye be o' Him."

At another time she said:

"I be all that hath been, and all that is, all that shalt be, for that be He."

Taken alone this would seem to be a declaration that she herself was God, but when it is read in connection with the previous affirmation it is readily understood.

"Thou art of Him," she said again, "aye, and I be of Him, and ye be of Him, and He be all and of all."

In this prayer, where she says "Teach me, O God, no word to pray," it is evident from her other prayers that she uses the word pray in the sense of "to beg." Her prayers are merely expressions of love and gratitude.

She herself interprets the line, "Teach me, O God, to waste not word," in this verse:

Speak ye a true tongue,
Or waste ye with words the Soul's song?
A damning evidence is with wasted words;
For need I prate to yonder star
When hunger fills the world wherein I dwell?
Cast I a glance so precious as His
Which wakes at every dawn?
Speak I a tongue one half so true
As sighing winds who sing amid
Aeolian harps strung with siren tress?
For lo, the sea murmureth a thousand tones,
Wrung from its world within,
But telleth only of Him,
And so His silence keeps.

In the order in which we have chosen to present these poems, they are more and more mystical as we go on. We trust, then, that the reader meeting them for the first time will feel no impertinence in increasing attempts at elucidation from one who has read them often and pondered them much.

There is another and a very interesting phase of these communications in the place Christ holds in them. Patience's attitude toward the Savior is one of deep and loving reverence.

"Didst thou then," she says, "with those drops so worth, buy the throbbing at thy memory set aflutter? And is this love of mine so freely thine by that same purchase, or do I love thee for thy love of me?

And do I, then, my father's tilling for love of Him, like thee to shed my blood and tears for reapers in an age to come, because He wills it so? God grant 'tis so!"

Nor does she hesitate to assert His divinity with definiteness.

"Think ye," she cries, "that He who doth send the earth aspin athrough the blue depth o' Heaven, be not a wonder-god who springeth up where'er He doth set a wish! Yea, then doth He to spring from out the dust a lily; so also doth He to breathe athin (within) the flesh, and come unto the earth, born from out flesh athout the touch o' man. 'Tis so, and from off the lute o' me hath song aflowed that be asweeted o' the blood o' Him that shed for thee and me."

And she puts the same assertion of His divine birth into this tribute to the Virgin:

Mary, mother, thou art the Spring
That flowereth, though nay man aplanteth thee.
Mary, mother, the song of thee
That lulled His dreams to come,
Sing them athrough the earth and bring
The hope of rest unto the day.

Mary, mother, from out the side of Him
That thou didst bear, aflowed the crimson tide
That doth to stain e'en unto this day
The tide of blood that ebbed the man
From out the flesh and left the God to be.

Mary, mother, wilt thou then leave me catch
These drops, that I do offer them as drink
Unto the brothers of the flesh of me of earth?
Mary, mother of the earth's loved!
Mary, bearer of the God!
Mary, that I might call thee of a name befitting thee,
I seek, I seek, I seek, and none
Doth offer it to me save this:
Mother! Mother! Mother of the Him;
The flesh that died for me.

Chapter 8

THE IDEAS ON IMMORTALITY

Earth! Earth, the mother of us all!
Aye, the mother of us all!
How loth, bow loth, like to a child we be,
to leave and seek 'mid dark!
—Patience Worth.

IF the personality of Patience Worth and the nature and quality of her literary productions are worthy of consideration as evidences of the truth of her claim to a spiritual existence, then in the sufficiency of the proof may be found an answer to the world-old question: Is there a life after death? To what extent the facts that have been presented in this narrative may be accepted as proof, is for the reader to determine. But Patience has not been content to reveal a strange personality and a unique literature; she has had much to say upon this question of immortality. There is more or less spiritual significance in nearly all of her poetry and in some of her prose, and while her references to the after life are usually veiled under figures of speech, they nevertheless give assurances of its existence. She makes it clear, however, that she is not permitted to reveal the nature of that life beyond the veil, but she goes as far apparently as she dares, in the repeated assertion, through metaphor and illustration, of its reality.

"My days," she cries, "I have scattered like autumn leaves, whirled by raging winds, and they have fallen in various crannies 'long the way. Blown to rest are the sunny spring kissed mornings of my youth, and with many a sigh did I blow the sobbing eves that melted into tear-washed night. Blow on, thou zephyr of this life, and let me throw the value of each day to thee. Blow, and spend thyself, till, tired, thou wilt croon thyself to sleep. Perchance this casting of my day may cease, and thou wilt turn anew unto thy blowing and reap the casting of the world.

"What then is a sigh? Ah, man may breathe a sorrow. Doth then the dumbness of his brother bar his sighing? Nay—and hark! The sea doth sigh, and yonder starry jasmine stirreth with a tremorous sigh; and morning's birth is greeted with the sighing of the world. For what? Ah, for that coming that shall fulfill the promise, and change the sighing to a singing, and loose the tongue of him whom God doth know and, fearful lest he tell His bidden mysteries, hath locked his lips."

And again she asks: "Needest thou see what God himself sealeth thine eyes to make thee know?" Meaning, undoubtedly, that only through the process of death can the soul be brought to an understanding of that other life; and she declares that even if we were shown, we could not comprehend. "If thou should'st see His face on morrow's break," she says, "'twould but start a wagging," a discussion. And she continues: "Ah, ope the tabernacle, but look thou not on high, for when the filmy veil shall fade away—ah, could'st thou but know that He who waits hath looked, aye looked, on thee, and thou hast looked on Him since time began!" This enigmatical utterance is in itself sufficient to start a "wagging," but Patience evidently feels that the solution is beyond our powers: for she repeatedly asserts that the key to the mystery is within our reach if we could but grasp it. "Fleet as down blown from its moorings, seeking the linnet who dropped her seed, so drift ye," she says, "ever seeking, when at the root still rests the seed pod." And again: "Knowest thou that fair land to which the traveler is loath to go, but loath, so loath, to leave? Ah, the mystery of the snail's shell is far deeper than this."

Yet she tells us again and again that Nature itself is the proof of another life. "Why live," she asks, "the paltry span of years allotted thee, in desolation, while all about thee are His promises? Thou art, indeed, like a withered band that holds a new-blown rose." The truth, she says, is not to be found in "books of wordy filling," but in the infant's smile and in the myriad creations and resurrections that are ever within our cognizance. "I pipe of learning," she cries, "and fall silent before the fool who singeth his folly lay."

The natural evidences she points out are visible to all and within the comprehension of the feeblest intelligence, but he whose vision is obscured by book knowledge "is like unto the monk who prays within his cell, unheedful of the timid sunbeam who would light the page his wisdom so befogs." "Ah!" she exclaims, "the labor set thee to unlearn thine inborn fancies!" meaning, apparently, the suppression of the intuitions of immortality; and in the same line of thought she cries: "Am I then drunkened on the chaff of knowledge supped by mine elderborn? Nay, my forefolk drank not truth, but sent through my veins acoursing, chaff, chaff, naught by chaff." Plainly, then, Patience has no great respect for learning, and it is the book of Nature rather than the book of words that she would have us read.

I made a song from the dead notes of His birds,
And wove a wreath of withered lily buds,
And gathered daisies that the sun bad scorched,
And plucked a rose the riotous wind had torn,
And stolen clover flowers, down-trodden by the kine,
And fashioned into ropes and tied with yellow reed,
An offering unto Him: and lo, the dust
Of crumbling blossoms fell to bloom again,
And smiled like sickened children,
Wistfully, but strong of faith that mother-stalk
Would send fresh blossoms in the spring.

So it is she sings, presenting the symbolisms of nature to illustrate the renewal or the continuance of life; or again, she likens life to the seasons (as did Shakespeare and Keats, and many another poet) in this manner:

My youth is promising as spring,
And verdant as young weeds,
Whose very impudence taketh them
Where bloom the garden's treasures.
My midlife, like the summer, who blazeth
As a fire of blasting beat, fed by withered
Crumbling weeds of my spring.
My sunset, like the fall who ripeneth
The season's offerings. And boar frost
Is my winter night, fraught with borrowed warmth,
And flowers, and filled with weeds,
Which spring e'en 'neath the frozen waste?

Ah, is the winter then my season's close?
Or will I pin a faith to hope and look
Again for spring, who lives eternal in my soul?

Faith is the keynote of many of her songs, the faith that grows out of that profound love which is the essential principle of the religion she presents. The triumph of faith she expresses in the poem which follows:

O sea! The panting bosom of the Earth;
The sighing, singing carol of her heart!
I watch thee and I dream a dream
Whose fruit doth sicken me.
White sails do fleck thy sheen, and yonder moon
Doth seem to dip thy depths
And sail the silver mirror, high above.
Unharbored do I rove. Along the shore behind,
The shadow of Tomorrow creepeth on.
A seething silvered path doth stretch thy length,
To meet the curving check of Lady Moon.
I dream the flutt'ring waves to fanning wings
And fain would follow in their course. But stay!
My barque doth plow anew, and set the wings to flight;
For though I watch their tremorous mass, my craft
But saileth harbor-loosed, and ever stretcheth far
Beyond the moon's own phantom path
And I but dream a dream whose fruit doth sicken me.
Ah, Sea! who planted thee, and cast
A silver purse, unloosed, upon thy breast?

My barque, who then did harbor it,
And who unfurled its sail?
And yonder moon, from whence her silver coaxed?
Methinks my dream doth wax her wroth,
Else why the pallor o'er her cast?
Dare I to sail, to steer me at the wheel?
Shall I then hide my face and cease my murmuring,
O'erfearful lest I find the port?
Nay, I do know thee, Lord, and fearless sail me on,
To harbor then at dawning of new day.
I stand unfearful at the prow.

At anchor rests my barque.
Away, thou phantom Moon,
And restless, seething path!
My chart I cast unto the sea,
For I do know Thee, Lord!

This triumph of faith is also the theme of the weird allegory which follows. It is, perhaps, the most mystical of Patience's productions.

The Phantom and the Dreamer

Phantom:
Thick stands the hill in garb of fir,
And winter-stripped the branching shrub.
Cold gray the sky, and glistered o'er
With star-dust pulsing tremorously.

Snow, the lady of the Winter Knight,
Hath danced her weary and fallen to her rest.
She lieth stretched in purity
And dimpled 'neath the trees.
A trackless waste doth lie from hill
To valley 'neath, and Winter's Knight
Doth sing a wooing lay unto his love.

Cot on cot doth stand deserted,
And thro' the purpled dark they show
Like phantoms of a life long passed
To nothingness. Hear thou the hollowness
Of the sea's coughing beat against
The cliff beneath, and harken ye
To the silence of the valley there.
Doth chafe ye of thy loneliness?
Then sleep and let me put a dream to thee.

See ye the cot—
A speck o' dark adown the hillside,
And sheltered o'er with fir-bows,
Heavy-laden with the kiss of Lady Snow?
Come hither then. Let's bruise this snowy breast,

And fetch us there unto its door.

See! Here a twig
Hath battled with the wind, and lost.
We then may cast it 'mid its brothers
Of the bush and plow us on.
Look ye to the thick thatch
O'er the gable of the roof,
Piled higher with a blanketing of snow;
And shutters hang agape, to rattle
Like the cackle of a crone.
The blackness of a pit within,
And filled with sounds that tho' they be
But seasoning of the log, doth freeze
Thy marrowmeat. I feel the quake
And shake thee for thy fear.

Stride thou within and set a flint to brush
Within the chimney-place.
We then shall rouse
The memory of the tenant here
A night, my friend, thee'lt often call to mind.
The flame hath sprung and lappeth at the twigs.
Thee'lt watch the burning of thy hastiness,
And wait thee long
Until the embers slip away to smoke. Then strain ye to its weaving
And spell to me the reading of its folds.

Dreamer:
I see thin, threading lines that writhe them
To a shape—a visage ever changeful,
Or mine eyes do play me false,
For it doth smile to twist it to a leer,
And sadden but to laugh in mockery.
I see a lad whose face
Doth shine illumed, and he doth bear
The kiss of wisdom on his brow.
I see him travail 'neath a weary load,
And close beside him Wisdom follows on.
Burdened not is be. Do I see aright?

For still the light of wisdom shineth o'er. But stay!
What! Do mine eyes then cheat?
This twisting smoke-wreath
Filleth all too much my sight!

Phantom:
Nay, friend, strain thee now anew.
The lad! Now canst thou see?
Nay, for like to him
Thou hast looked thee at the face of Doubt.

Dreamer:
Who art thou, shape or phantom, then,
That thou canst set my dream to flight?
I doubt me that the lad could stand
Beneath the load!

Phantom:
Nay, thee canst ravel well, my friend.
The lad was thee, and Doubt
O'ertook with Wisdom on thy way.
Come, bury Doubt aneath the ash.
We travel us anew.
Seest thou, a rimming moon doth show
From 'neath the world's beshadowed side.
A night bird chatteth to its mate,
And lazily the fir-boughs wave.
We track us to the cot whose roof
Doth sag—and why thy shambling tread?
I bid ye on!

Dreamer:
Who art thou—again I that demand—
That I shall follow at thy bidding?
Who set me then this task?

Phantom:
Step thou within!
Stand thee on the thresh of this roofless void!
Look thou! Dost see the maid

Who coyly stretcheth forth her band
To welcome thee? She biddeth thee
To sit and sup. I bid thee speak.
Awaken thee unto her welcoming.

Dreamer:
Enough! This fancy-breeding sickeneth
My very soul! A skeleton of murdered trees,
Ribbed with pine and shanked of birch!
And thee wouldst bid me then
Embrace the emptiness.
I see naught, and believe but what I see.

Phantom:
Look thou again, and strain.
What seest thou?

Dreamer:
I see a newly kindled fire,
And watch its burning glow until
The embers die and send their ghosts aloft.
But ash remaineth—and I chill! For rising there, a shape
Whose visage twisteth drunkenly,
And from her garments falls a dust of ash."

Phantom:
Doubt! Unburied, friende! We journey on,
And mark ye well each plodding footfall
Singing like to golden metal with the frost.
The night a scroll of white, and lined
With blackish script—
The lines of His own putting!
Read thee there! Thou seest naught,
And believe but what ye see!
Stark nakedness and waste but hearken ye!
The frost skirt traileth o'er the crusted snow
And singeth young leaves' songs of Spring.

Still art thou blind!
But at His touching shall the darkness bud

And bloom to rosy morn. And even now,
Were I to snap a twig 'twould bleed and die.
See ye; 'tis done! Look ye!
Ye believe but what ye see:
Here within thy very band
Thou boldest Doubt's undoing.
I bid ye look upon the bud
Already gathered 'neath the tender bark.
The sun's set and rise hath coaxed it forth.
Thee canst see the rogue hath stolen red
And put it to its heart. And here
Aneath the snow the grass doth love the earth
And nestles to her breast.
I stand me here, and lo, the Spring hath broke!
The dark doth slip away to bide,
And flowering, singing, sighing, loving
Spring Is here!

Dreamer:
Aye, thou art indeed
A wonder-worker in the night!
A black pall, a freezing blast,
An unbroken path—and thou
Wouldst have me then to prate o' Spring,
And pluck a bud where dark doth bide the bush!
Who cometh from the thicket higher there?

Phantom:
'Tis Doubt to meet thee, friend!

Dreamer:
Who art thou? I fain would flee,
And yet I fear to leave lest I be lost.
I hate thee and thy weary task!

Phantom:
Nay, brother, thy lips do spell,
But couldst thee read their words aright
Thee wouldst meet again with Doubt.
Come! We journey on unto the cot

Beloved the most by me. I bid thee
Let thy heart to warm within thy breast.
A thawing melteth frozen Hope.
See bow, below, the sea hath veiled
Her secret held so close,
And murmured only to the winds
Who woo her ever and anon.
The waves do lap them, hungry for the sands.
Careful! Lest the sun's pale rise
Should blind thee with its light.
A shaft to put it through
The darkness of thy soul must needs
But be a glimmering to blind.
Step ye to the hearthstone then,
And set thee there a flame anew.
I bid ye read again
The folding of the smoke.

Dreamer:
'Tis done, thou fiend!
A pretty play for fools, indeed.
I swear me that 'tis not
For loving of the task I builded it,
But for the warming of its glow.

Phantom:
In truth ye speak. But read!

Dreamer:
I see a hag whose brow O
Doth wrinkle like a summer sea.
For do I look unto the sea
At Beauty's own fair form,
It writheth to a twisted shape,
And I do doubt me of her loveliness.
The haggard visage of the crone
I now behold, doth set me doubting
Of mine eye, for dimples seem
To flutter 'neath the wrinkled cheek.

Phantom:
So, then, thee believest
But what thine eyes behold!
Thee findest then
Thy seeing in a sorry plight.
I marvel at thy wisdom, lad.
Look ye anew.
Mayhap thee then
Canst coax the crone away.

Dreamer:
Enough! The morn hath kissed the night adieu,
And even while I prate
A redwing crimsoneth the snow in flight.
Kindled tinder smoldereth away,
And I do strain me to its fold.
I glut me of the loveliness I there behold,
For from the writhing stream a sprite is born
Whose beauteous form bedazzles me,
And she doth point me
To the golding gray of morn. The sea
Is singing, singing her unto my soul.
I dreamed she sighed, but waked to hear her sing.
I hear thee, Phantom, bidding me on, on!
But morn hath stolen dreams away.
I strain me to the hills to trace our path,
And lo, unbroken is the snow,
And cots have melted with the light,
And yet, methinks a murmuring doth come
From out the echoes of the night,
That bid them 'neath the crannies of the hills.
Life! Life! I lead thee on!
And faith doth spring from seedlings of thy doubt!

EPILOGUE

THICK stands the hill in garb of fir and snow.
The Lady of the Winter's Knight hath danced
Her weary, and stretched her in her purity,
To cover aching wounds of Winter's overloving woo.

"And faith doth spring from seedlings of thy doubt!" plainly meaning an active doubt that searches for the truth and finds it. But she personifies Doubt in another and more forbidding form in this:

Like to a thief who wrappeth him
Within the night-tide's robe,
So standeth the specter o' the Earth;
Yea, he doth robe him o' the Earth's fair store.
Yea, he decketh in the star-hung purple o' the eve,
And reacheth from out the night unto the morn,
And wringeth from her waking all her gold,
And at his touching, lo, the stars are dust,
And morn's gold but heat's glow, and ne'er
The golden blush of His own metal store.

Yea, be strideth then
Upon the flower-hung couches of the field,
And traileth him thereon his robe,
And lo, the flowers do die of thirst
And parch of scoarching of his breath.

Yea, and 'mid the musics of the earth he strideth him,
And full-songed throats are mute.
Yea, music dieth of his luring glance.
And e'en the love of earth he seeketh out
And turneth it unto a folly-play.
Yea, beneath his glance, the fairy frost
Upon the love sprite's wing
Doth flutter, as a dust, and drop, and leave
But bruised and broken bearers for His store.

Yea, and 'mid man's day he ever strideth him
And layeth low man's reasoning. His robes
Are hung of all the earth's most loved.
From off the flowers their fresh; from off the day
The fairness of her hours. For dark, and bid
Beneath his cloak, he steppeth ever,
And doth hiss his name to thee
Doubt.

I have said that the message of Patience Worth contained a revelation, a religion and a promise. The revelation is too obvious to need a pointer. In the preceding chapter were presented the elements of the religion that she reveals, with which should be included the unfaltering faith expressed in these poems. Love and Faith—these are the two Graces upon whom, to personify them, all her work is rested, and from them spring the promise she conveys. That promise has to do with the hereafter, and Patience knows the human attitude in relation to that universal problem, and she gives courage to the shrinking heart in this poem on the fear of death:

I stride abroad before my brothers like a roaring lion,
Yet at even's close from whence cometh the icy band
That clutcheth at my heart and maketh me afraid—
The slipping of myself away, I know not whither?
And lo, I fall atremble.
When I would grasp a straw, 'tis then I find it not.
Can I then trust me on this journey lone
To country I deem peopled, but know not?
My very heart declareth faith, yet hath not thine
Been touched and chilled by this same phantom?

Ah, through the granite sips the lichen
And hast thou not a long dark journey made?
Why fear? As cloud wreaths fade
From spring's warm smile, so shall fear
Be put to flight by faith.

I pluck me buds of varied hue and choose the violet
To weave a garland for my loved and best.
I search for bloom among the rocks
And find but feathery plume.
I weave, and lo, the blossoms fade
Before I reach the end,
And faded lie amid my tears—
And yet I weave and weave.
I search for jewels 'neath the earth,
And find them at the dawn,
Besprinkled o'er the rose and leaf,
And showered by the sparrow's wing,
Who seeketh 'mid the dew-wet vine
A harbor for her home.
I search for truth along the way
And find but dust and web,
And in the smile of infant lips
I know myself betrayed.
I watch the swallow skim across the blue
To homelands of the South,
And ah, the gnawing at my heart doth cease;
For bow be wings and wings
To lands be deemeth peopled by his brothers,
Whose song be hears in flight!
Not skimming on the lake's fair breast is be,
But winging on and on,
And dim against the feathery cloud
He fades into the blue.

I stand with withered blossoms crushed,
And weave and weave and weave.

This is Patience's answer to the eternal question:

Can I then trust me on this journey lone
To country I deem peopled, but know not?

It is the cry of him who believes and yet doubts, and Patience points to the swallow winging across the blue "to lands be deemeth peopled with his brothers" who have gone on before. In imagination he can hear their song in the home lands of the South, and though he cannot see them, and cannot have had word from them, he knows they are there, and he does not skim uncertainly about the lake, but with unfaltering faith "wings him on and on" until—

Dim against the feathery cloud
He fades into the blue.

But Patience does not content herself with appeals to faith, eloquent as they may be.

While her communications are always clothed in figures of speech, they are sometimes more definite in statement than in the lines which have been thus far presented. In the prose poem which follows, she asks and answers the question in a way that can leave no doubt of her meaning:

"Shall I arise and know thee, brother, when like a bubble I am blown into Eternity from this pipe of clay? Or shall I burst and float my atoms in a joyous spray at the first beholding of this home prepared for thee and me, and shall we together mingle our joys in one supreme joy in Him? It matters not, beloved, so comfort thee. For should the blowing be the end, what then? Hath not thy pack been full, and mine? We are o'erweary with the work of living, and sinking to oblivion would be rest. Yet sure as sun shall rise, my dust shall be unloosed, and blow into new fields of new days. I see full fields yet to be harvested, and I am weary. I see fresh business of living, work yet to be done, and I am weary. Oh, let me fold these tired hands and sleep. Beloved, I trust, and expect my trust, for ne'er yet did He fail."

She puts this into the mouth of one who lives, but it is not merely an expression of faith; it is a positive assertion. "Yet sure as sun shall rise, my dust shall be unloosed, and blow into new fields of new days."

And again she sings:

What carest, dear, should sorrow trace
Where dimples sat, and should
Her dove-gray cloud to settle 'neath thine eye?
The withering of thy curving cheek
Bespeaks the spending of thy heart.
Lips once full are bruised
By biting of restraint. Wax wiser, dear.
To wane is but to rest and rise once more.

Or she puts the thought in another form in this assurance:

Weary not, O brother!
'Tis apaled, the sun's gold sink.
Then weary not, but set thy path to end,
E'en as the light doth fade and leave
Nay trace to mar the night's dark tide.
Sink thou, then, as doth the sun,
Assured that thou shalt rise!

All these, however, are but preparatory to the communication in which she asserts not only the actuality of the future life but something of the nature of it. One might say that the preceding poems and prose—poems, taken alone and without regard to the mystery of their source, were merely expressions of belief, but in this communication she seems to speak with knowledge, seems even to have overstepped the bounds within which, she has often asserted, she is held. "My lips be astopped," she has said in answer to a request for information of this forbidden character, but here she appears to have been permitted to give a glimpse of the unknown, and to present a promise of universal application. This poem, from the spiritual standpoint, is the most remarkable of all her productions.

How have I caught at fleeting joys
And swifter fleeting sorrows!

And days and nights, and morns and eves,
And seasons, too, aslipping thro' the years, afleet.
And whither hath their trend then led?
Ah, whither!

How do I to stop amid the very pulse o' life.
Afeared! Yea, fear clutcheth at my very heart!
For what? The night? Nay, night doth shimmer
And flash the jewels I did count
E'er fear bad stricken me.

The morn? Nay, I waked with morn atremor,
And know the day-tide's every hour.
How do I then to clutch me
At my heart, afeared?
The morrow? Nay,
The morrow but bringeth old loves
And hopes anew.

Ah, woe is me, 'tis emptiness, aye, naught—
The bottomlessness o' the pit that doth afright!
Afeared? Aye, but driven fearless on!

What! Promise ye 'tis to mart I plod?
What! Promise ye new joys?
Ah, but should I sleep, to waken me
To joys I ne'er had supped!

I see me stand abashed and timid,
As a child who cast a toy beloved,
For bauble that but caught the eye
And left the heart ahungered.

What! Should I search in vain
To find a sorrow that had fleeted hence
Afore my coming and found it not?
Ah, me, the emptiness!

And what! should joys that but a prick
Of gladness dealt, and teased my hours
To happiness, be lost amid this promised bliss?
Nay, I clutch me to my heart In fear, in truth!

Do harken Ye! And cast afearing
To the wiles of beating gales and wooing breeze.

I find me throat aswell and voice attuned.
Ah, let me then to sing, for joy consumeth me!
I've builded me a land, my mart,
And fear hath slipped away to leave me sing.
I sleep, and feel afloating.
Whither! Whither! To wake,—
And wonder warmeth at my heart,
I've waked in yester-year!

What! Ye? And what! I'st thou?
Ah, have I then slept, to dream? Come,
Ne'er a dream-wraith looked me such a welcoming!
'Twas yesterday this band wert then afold,
And now,—ah, do I dream?
'Tis warm-pressed within mine own!
Dreams! Dreams! And yet, we've met afore!

I see me flitting thro' this vale,
And tho' I strive to spell
The mountain's height and valley's depth,
I do but fall afail.
Wouldst thou then drink a potion
Were I to offer thee an empty cup?
Couldst thou to pluck the rainbow from the sky?
As well, then, might I spell to thee.

But I do promise at the waking,
Old joys, and sorrows ripened to a mellow heart.
And e'en the crime-stained wretch, abasked in light,
Shall cast his seed and spring afruit!

Then do I cease to clutch the emptiness
And sleep, and sleep me unafeared!

What is it that affrights, she asks, when we think of death?

It is the emptiness, she answers, the utter lack of knowledge of what lies beyond. And if we waken to "joys we ne'er have supped"—using the word sup in the sense of to taste or to know—what is there to attract us in the prospect? It is an illustration she presents of our attitude toward promises of joys with which we are unfamiliar; and which therefore do not greatly interest us— the child who casts aside a well beloved toy "for bauble that but caught the eye and left the heart ahungered." Shall the joys, she makes us exclaim, which we have known here but barely tasted in this fleeting life, "be lost amid this promised bliss!" and shall we "search in vain to find a sorrow that bad fleeted hence before our coming?"—meaning, apparently, shall we look there in vain for a loved one who has gone before? She answers these questions of the heart. Personality persists beyond the grave, she gives us plainly to understand. We take with us all of ourselves but the material elements. "Thou art ye," she has said, "and I be me and ye be ye, aye, ever so. The transition is but a change from the material to the spiritual. We "wake in yesteryear," she says,—amid the friends and associations of the past; and the joys of that life, one must infer, are the spiritual joys of this one, the joy that comes from love, from good deeds, from work accomplished. For it is quite evident that she would have us believe that there is a continuous advancement in that other life.

And e'en the crime-stained wretch, abasked in light,
Shall cast his seed and spring afruit.

This can mean nothing else than that the hardened sinner, amid supernal influences, shall develop into something higher, and as no one can be supposed to be perfect when leaving earth, it follows that progress is common to all. Progress implies effort, and this indicates that there will be something for everyone to do—a view quite different from the monotony of eternal idleness.

But this I promise at the waking,
Old joys, and sorrows ripened to a mellow heart.

To those who would peer into the other land these are perhaps the most important lines she has given. But what does she mean by "sorrows ripened to a mellow heart?" She was asked to make that plainer and she said:

"That that hath flitted hence be sorrows of earth, and ahere be ripened and thine. Love alost be sorrow of earth and dwell ahere."

She thus makes these lines an answer to the question put before:

What! Should I search in vain
To find a sorrow that bad fleeted hence
Afore my coming and found it not?

These are the sorrows that are "ripened to a mellow heart," and she was asked if there were new sorrows to be borne in that other life. She replied:

"Nay. Earth be a home of sorrow's dream.
For sorrow be but dream of the soul asleep.
'Tis wake (death) that setteth free."

And after such assurance combs the cry of faith and content and peace:

Then do I cease to clutch the emptiness,
And sleep, and sleep me unafeared!

With this comforting assurance in mind one may cheerfully approach her solemn address to Death:

Who art thou,
Who tracketh 'pon the path o' me
O' each turn, aye, and track?

Thou! And thou astand!
And o'er thy face a cloud,
Aye, a darked and somber cloud!
Who art thou,
Thou tracker 'mid the day's bright,
And 'mid the night's deep;
E'en when I be astopped o' track?

Who art thou,
That toucheth o' the flesh o' me,
And sendeth chill unto the heart o' me?
Aye, and who art thou,
Who putteth forth thy hand
And setteth at alow the hopes o' me?

Aye, who art thou,
Who bideth ever 'mid a dream?

Aye, and that the soul o' me
Doth shrink at know?

Who art thou? Who art thou,
Who steppeth ever to my day,
And blotteth o' the sun away?

Who art thou,
Who stepped to Earth at birth o' me,
And e'en 'mid wail o' weak,
Aye, at the birth o' wail,
Did set a chill 'pon infant flesh;
And at the track o' man 'pon Earth
Doth follow ever, and at height afollow,
And doth touch,
And all doth crumble to a naught.
Thou! Thou! Who art thou?
Ever do I to ask, and ever wish
To see the face o' thee,
And ne'er, ne'er do I to know thee
Thou, the Traveler 'pon the path o' me.
And, Brother, thou dost give
That which world doth hold
From see o' me!

Stand thou! Stand thou!
And draw thy cloak from o'er thy face!

Ever hath the dread o' thee
Clutched at the heart o' me.
Aye, and at the end o' journey,
I beseech thee,
Cast thy cloak and show thee me!
Aye, show thee me!

Ah, thou art the gift o' Him!
The Key to There! The Love o' Earth!

Aye, and Hate hath made o' man
To know thee not—
Thou! Thou! O Death!

She finds Death terrible from the human point of view, and reveals him at the end as "the gift of Him, the Key to There!" One of her constant objects seems to be to rob death of its terrors, and to bring the "There" into closer and more intimate connection with us. Here is another effort:

Spring's morn afulled o' merry-song,
Aye, and tickle o' streams-thread through Summer's noon;

Arock o' bum o' hearts-throb,
And danced awhite the air at scorch;

Winter's rage asing o' cold
And wail o' Winter's sorry at the Summer's leave;

Ashivered breeze, abear o' leaf's rustling
At dry o' season's ripe;

Night's deep, where sound astarteth silence;
Morn's sweet, awooed by bird's coax.

Earth's sounds, ye deem?
I tell thee 'tis but the echoing o' Here.

Thy days be naught
Save coax o' Here athere!

All that is worthwhile on earth is but the echoes of Heaven, and there would be nothing to life but for the joys that have been "coaxed" from there. How closely that thought unites the here and the there. Earth sounds but the echoes of the other land adjoining! She makes it something tangible, something almost material, something we may nearly comprehend; and then, having opened the door a little way, as far, no doubt, as it is possible for her to do, she presents this response to human desires, this promise of joys to come:

Swift as light-flash o' storm, swift, swift,
Would I send the wish o' thine asearch.
Swift, swift as bruise o' swallows' wing 'pon air,
I'd send asearch thy wish, areach to lands unseen;
I'd send aback o' answer laden.
Swift, swift, would I to flee unto the Naught
Thou knowest as the Here.
Swift, swift I'd bear aback to thee\

What thou wouldst seek. Swift, swift,
Would I to bear aback to thee.
Dost deem the path ahid doth lead to naught?
Dost deem thy footfall leadest thee to nothingness?
Dost pin not 'pon His word o' promising,
And art at sorry and afear to follow Him?
I'd put athin thy cup a sweet, a pledge o' love's-buy.
I'd send aback a glad-song o' this land.
Sing thou, sing on, though thou art ne'er aheard—
Like love awaked, the joy o' breath
Anew born o' His loving.

Set thee at rest, and trod the path unfearing.
For He who putteth joy to earth, aplanted joy
Athin the reach o' thee, e'en through
The dark o' path at end o' journey.
His smile! His word! His loving!
Put forth thy band at glad, and I do promise thee
That Joy o' earth asupped shall fall as naught,
And thou shalt sup thee deep o' joys,
O' Bearer, aye, and Source; and like glad light o' day
And sweet o' love, thy coming here shall be!

With this promise, this covenant, we bring the narrative of Patience to an end. There will be many and widely varied views of the nature of this intelligence, but surely there can be but one opinion of the beauty of her words and the purity of her purpose. She has brought a message of love at a time when the world is sadly deficient in that attribute, wisely believed to be the best thing in earth or heaven; and an inspiration to faith that was never so greatly in need of strength as now. An inevitable consequence of the world-war will be a universal introspection. There will be a great turning of thought to serious things. That tendency is already discernible. May it not be possible that it is the mission of Patience Worth to answer the question that is above all questions at a time when humanity is filled with interrogation?

FINIS.

INDEX

Paperbacks also available from White Crow Books

Marcus Aurelius—*Meditations*
ISBN 978-1-907355-20-2

Elsa Barker—*Letters from a Living Dead Man*
ISBN 978-1-907355-83-7

Elsa Barker—*War Letters from the Living Dead Man*
ISBN 978-1-907355-85-1

Elsa Barker—*Last Letters from the Living Dead Man*
ISBN 978-1-907355-87-5

Richard Maurice Bucke—*Cosmic Consciousness*
ISBN 978-1-907355-10-3

G. K. Chesterton—*Heretics*
ISBN 978-1-907355-02-8

G. K. Chesterton—*Orthodoxy*
ISBN 978-1-907355-01-1

Arthur Conan Doyle—*The Edge of the Unknown*
ISBN 978-1-907355-14-1

Arthur Conan Doyle—*The New Revelation*
ISBN 978-1-907355-12-7

Arthur Conan Doyle—*The Vital Message*
ISBN 978-1-907355-13-4

Arthur Conan Doyle with Simon Parke—*Conversations with Arthur Conan Doyle*
ISBN 978-1-907355-80-6

Meister Eckhart with Simon Parke—*Conversations with Meister Eckhart*
ISBN 978-1-907355-18-9

Kahlil Gibran—*The Forerunner*
ISBN 978-1-907355-06-6

Kahlil Gibran—*The Madman*
ISBN 978-1-907355-05-9

Kahlil Gibran—*The Prophet*
ISBN 978-1-907355-04-2

Kahlil Gibran—*Jesus the Son of Man*
ISBN 978-1-907355-08-0

Kahlil Gibran—*Spiritual World*
ISBN 978-1-907355-09-7

D. D. Home—*Incidents in my Life Part 1*
ISBN 978-1-907355-15-8

Mme. Dunglas Home; edited, with an Introduction, by Sir Arthur Conan Doyle—*D. D. Home: His Life and Mission*
ISBN 978-1-907355-16-5

Edward C. Randall—*Frontiers of the Afterlife*
ISBN 978-1-907355-30-1

Lucius Annaeus Seneca—*On Benefits*
ISBN 978-1-907355-19-6

Rebecca Ruter Springer—*Intra Muros: My Dream of Heaven*
ISBN 978-1-907355-11-0

Leo Tolstoy, edited by Simon Parke—*Forbidden Words*
ISBN 978-1-907355-00-4

Leo Tolstoy—*A Confession*
ISBN 978-1-907355-24-0

Leo Tolstoy—*The Gospel in Brief*
ISBN 978-1-907355-22-6

Leo Tolstoy—*The Kingdom of God is Within You*
ISBN 978-1-907355-27-1

Leo Tolstoy—*My Religion: What I Believe*
ISBN 978-1-907355-23-3

Leo Tolstoy—*On Life*
ISBN 978-1-907355-91-2

Leo Tolstoy—*Twenty-three Tales*
ISBN 978-1-907355-29-5

Leo Tolstoy—*What is Religion and other writings*
ISBN 978-1-907355-28-8

Leo Tolstoy—*Work While Ye Have the Light*
ISBN 978-1-907355-26-4

Leo Tolstoy with Simon Parke—*Conversations with Tolstoy*
ISBN 978-1-907355-25-7

Vincent Van Gogh with Simon Parke—*Conversations with Van Gogh*
ISBN 978-1-907355-95-0

Howard Williams with an Introduction by Leo Tolstoy—*The Ethics of Diet: An Anthology of Vegetarian Thought*
ISBN 978-1-907355-21-9

Allan Kardec—*The Spirits Book*
ISBN 978-1-907355-98-1

Wolfgang Amadeus Mozart with Simon Parke—*Conversations with Mozart*
ISBN 978-1-907661-38-9

Jesus of Nazareth with Simon Parke—*Conversations with Jesus of Nazareth*
ISBN 978-1-907661-41-9

Thomas à Kempis with Simon Parke—*The Imitation of Christ*
ISBN 978-1-907661-58-7

Emanuel Swedenborg—*Heaven and Hell*
ISBN 978-1-907661-55-6

P.D. Ouspensky—*Tertium Organum: The Third Canon of Thought*
ISBN 978-1-907661-47-1

Dwight Goddard—*A Buddhist Bible*
ISBN 978-1-907661-44-0

Leo Tolstoy—*The Death of Ivan Ilyich*
ISBN 978-1-907661-10-5

Leo Tolstoy—*Resurrection*
ISBN 978-1-907661-09-9

Michael Tymn—*The Afterlife Revealed*
ISBN 978-1-970661-90-7

Guy L. Playfair—*If This Be Magic*
ISBN 978-1-907661-84-6

Julian of Norwich with Simon Parke—*Revelations of Divine Love*
ISBN 978-1-907661-88-4

Maurice Nicoll—*The New Man*
ISBN 978-1-907661-86-0

Carl Wickland, M.D.—*Thirty Years Among the Dead*
ISBN 978-1-907661-72-3

Allan Kardec—*The Book on Mediums*
ISBN 978-1-907661-75-4

John E. Mack—*Passport to the Cosmos*
ISBN 978-1-907661-81-5

All titles available as eBooks, and selected titles available in Hardback and Audiobook formats from www.whitecrowbooks.com